THE OWNER-BUILT
HOMESTEAD

THE OWNER–BUILT HOMESTEAD

Barbara and Ken Kern

CHARLES SCRIBNER'S SONS   NEW YORK

Library of Congress Cataloging in Publication Data

Kern, Ken
  The owner-built homestead.

  Bibliography: p. 355
  Includes index.
  1. Agriculture—Handbooks, manuals, etc. 2. Agriculture—United
States—Handbooks, manuals, etc.
I. Kern, Barbara, joint author. II. Title.
S501.2.K47   630'.973   77-4032
ISBN 0-684-14922-2
ISBN 0-684-14926-5 pbk.

1 3 5 7 9 11 13 15 17 19 Q/C 20 18 16 14 12 10 8 6 4 2
1 3 5 7 9 11 13 15 17 19 Q/P 20 18 16 14 12 10 8 6 4 2

Printed in the United States of America

To Mildred Loomis, Education Director of the School of Living, who, in 1947, supplied us with the first inspiration and example for this work; and to Doe Coover, editor at Charles Scribner's Sons, who, in 1977, diligently supplied the final polish and coherence to this work.

CONTENTS

Foreword

This book is a source of direction to the relatively small part of the populace that is aware of the untenability of our entire life situation. Inasmuch as this is so, *The Owner-Built Homestead* does not deal with the present but with the future. It is a starting point for systematic experimentation. It brings together and sorts out enough information on individual topics to become a viable base for future study or, more important, a support for immediate action. It does this so that those who have developed the means to do something new, something based on the information generated by the doing of things, will have a starting point.

In *The Owner-Built Homestead* you will encounter a large amount of reportage on current practices and discussion of the many ills that we have brought upon ourselves with our modern methods. You may ask, What has all this to do with my 30 acres and my, perhaps escapist, life in the country? But the answer must be clear: If yours is the escapist life, this book has nothing to do with it. And, if not, then you must ask yourself why in the first place you have come by this 30 acres and what difference your custodianship of this bit of the earth is going to make to it. What life will this bit of land know that it might not have known were it married to another husbandman? And if you are like most of us in asking this question, you will discover that you know little of giving to the land and much of taking from it. And is this not the very basic problem that this book addresses—very little giving and very much taking?

There was a not-too-distant time when those who spoke of oil depletion apart from allowances were considered fools. There was a time when water, air, and, indeed, the entire surface of the planet were considered inexhaustible and ever-renewable and those who spoke otherwise were undoubtedly foolish. Yet don't we tend to consider water and air pollution and soil depletion in their narrowest sense? That is, don't we really tend to view them as localized problems—as *an* air problem, *a* water problem, *an* oil problem? We fool ourselves with the notion that there exist individual solutions to these problems: a new kind of engine, a better smokestack, or a cleaner nuclear reactor. We hear this all the time and we engage constantly in this sort of thinking. We are always about the "improving" of a single situation that is ultimately untenable. All the while we are coming to this realization, we glimpse to our dismay that the momentum created by our shortsightedness will not and cannot be brought to any sudden halt. In the shock induced by this furtive view of our reality we lapse into thoughts of reform but fail to grasp that we are confronted with the gradual, eventual destruction of an entire biosphere, a destruction more awful and unfathomable than the specter of the atom bomb.

To say this is quite unfashionable. But when we read of the rape and pillage of resources, often heretofore shielded from public awareness, who will there be to point out that this circumstance is real and that in some distant decade the paper shortage will be eclipsed by a soil shortage? (And what cartel will be organized then to profit from that catastrophe?) Why must the fact that a problem exists always emerge just when it is too late? Why must we always ask ourselves what is to be done when there is no longer anything that can be done?

The Owner-Built Homestead is one attempt, not at reform, but at experimentation with a view to the possibility of a new and viable mode of living on this planet. To achieve such viability, a new mode must be able to cope with present dangers if it is to shape a future where possibilities for life again accrue rather than diminish. And without an emotional as well as an intellectual understanding, all back-to-the-land motion will ultimately lead back to the old grind. This book is, therefore, a proposal for a frontier, and for this very reason it is full of ideas, full of enthusiasm. It is the journal of travels yet to be undertaken. And those who see this will use this book to outfit themselves for the trek.

Joe Bruno

Preface

Our frontispiece is taken from a nineteenth-century painting of unknown origin. It features a horse-drawn plow, either abandoned in the field or, more likely, at rest during evening hours. This implement, in its lonely furrow, symbolizes for us the end of the homestead tradition, seemingly reached in the last fifty years of swift progress by agribusiness. It may, however, symbolize a historic turning point, one from which the homestead movement will yet evolve into a human tool for sane living by the family unit.

In this book we do not advocate a return to traditional farming practice; in fact, much of our book speaks against the plow culture and modern tillage. Ours is an alternative, low-technology approach, based on an understanding of the growth process, the need for intensive food production, and the enduring value of the homestead way of life.

The germ of this book was formed during the late 1940s and the decade of the 1950s. Much of the material found here appeared at that time in a newsletter, *New Technic: Productive Living Research Notes,* sent out by Ken, who wrote, mimeographed, and mailed this paper to practicing and prospective homesteaders across the country. At a later time *New Technic* appeared with innovative design aids emanating from Ken's Home and Homestead Design Service.

Some years later *New Technic* was expanded, updated, and published in monthly installments by the *Mother Earth*

News. This series, called *The Owner-Built Homestead,* was subsequently printed in loose-leaf book form with this title.

Jacek Galazka, Director of Publishing for Charles Scribner's Sons, discovered the series and approached Ken with the prospect of republishing the material in a single, unified volume. Ken's first book, *The Owner-Built Home,* was at that time more complete and so was published first.

Putting this work in presentable order for the book market was a task of large order, requiring the teamwork that had begun in the early 1960s with our association. Over the years we had separately and jointly accumulated countless facts from numerous, often obscure sources and from our heritage and experience. Together we compiled file cabinets of correspondence and critical feedback, drawing from our own intimate homesteading experience as we raised our family. We also drew upon our homestead plans and designs for others.

To make a righteous job of this work, we enlisted the aid, common sense, and goodwill of homesteader friends, Joe and Susan Bruno of Camden, New York. While deeply involved in their own homestead-family development, they took time to read the new manuscript, edit it, and help with elements of the rewriting. Theirs was a generous investment and dedicated motivation.

Finally, we wish to thank all of our fellow homesteaders—correspondents and clients alike—for their advice, for their technical feedback, and for sharing an opportunity to experiment with new design concepts.

Barbara and Ken Kern
North Fork, California

Introduction

. . . whoever could make two ears of corn or two blades of grass to grow upon a spot of ground where only one grew before, would deserve better of mankind, and do more essential service to his country, than the whole race of politicians put together.   *Jonathan Swift,* Gulliver's Travels

In just the last few years the book trade in the United States has witnessed a proliferation of how-to books and manuals about homesteading. Many new titles conspicuously fill library shelves. And with all of this verbiage accumulating in a heretofore limited field of writing, the reader may question the need or advisability of preparing yet another homesteading book. But we would like to think that our book offers more than a rehash of Depression-style farming practice. This is a book with a difference, for we regard neohomesteading as the means by which people can create the most wholesome, well-balanced, multifarious, and aesthetically satisfying life for themselves. In 1934 Professor John Gifford of the University of Florida said of homesteading, in a small book on diversified tree-crop farming:

The furtherance of the . . . subsistence homestead . . . seems to me about the most essential thing that can give life and comfort to the majority of our people; in fact, the only permanent way out of the difficulties which beset the world. The small farm home is the essential basic unit of society. The prosperity and strength of any country can be measured by the number of small self-supporting homesteads it contains. The best nations of the world are not those with the greatest natural resources but with the largest number of small, self-supportive, free-of-debt homesites.

With this we wholeheartedly concur.

Neohomesteading involves more than just food production. It means a particular concern for land use and for fair and wise land apportionment as well. In this regard Thomas Jefferson was the first American agriculturist to recognize the need for equitable land distribution. In a letter to James Madison in 1789 Jefferson stated the case: "Whenever there is in any country uncultivated lands and unemployed poor, it is clear that the laws of property have been so far extended as to violate natural right." Today in America only a quarter of privately owned land is cultivated, and few of the remaining billion uncultivated acres are available to the unemployed poor.

The Food and Agriculture Organization (FAO) of the United Nations warns that the amount of land available for food production is diminishing globally. In 1920 the available productive land was 2 acres per world inhabitant. Today the FAO tells us that half of the world's people are malnourished or starving. We need immediately to increase the world food supply by 200 percent to bring about even minimal improvement in the present condition of human nutrition on this planet.

Malthusian economists, predicting worldwide starvation, claim that it is population increase, not restricted access to land, that will bring about this destruction of human life, and they therefore tout zero population growth as our only salvation from a disaster caused by limited world food supply. This panacea, however, is more in the corporate interest, which, the world over, yearly engulfs vast agricultural territories with multi-square-mile factory farms. What the food business · really fears is an expanding world population that would create a revolutionary clamor for the land that could, indeed, be cultivated—if only it were available to people who need it and who would work it for their sustenance.

According to UN figures, the 3.5 billion of the world's acres now under cultivation could be increased to as much as 15 billion productive acres and could conceivably support a world population of as many as 500 billion people, or about 150 times our present population. Demographers' charts depicting the impact of rapid population increase certainly contradict this.

The anarchist Peter Kropotkin was the first seriously to challenge Malthusian economic philosophy. In the late nineteenth century Kropotkin, in *Fields, Factories and Workshops,* maintained that agriculture had barely begun to realize its potential. He therefore proposed that intensive farming be practiced on small parcels of land. The present policy of agribusiness is, however, based on extensive equipment farming. Plentiful land is available to the world's wealthy, and they work it with minimal use of labor. As a result, land is exploited for the greatest economic return to the corporations that control it, even though the result is often a low per-acre crop yield.

Today, a hundred years after Kropotkin's revelations about agriculture's potential, the consensus among agricultural economists and even corporate farm management is that "the most efficient producing unit is a farm that can be run by its owner" (*Fortune,* August 1972). Even with generous tax write-offs and crop subsidy welfare, agribusiness farming is far less efficient than owner-operated enterprise. Yet the trend continues for big-farm expansion. One suspects—justifiably—that this is the result of the regrettable fact that America's primary exportable political asset is its food basket, which, paradoxically, rarely reaches the hungry and starving anywhere.

The 5.5 million American farms of the 1950s were by 1976 reduced to half that number. The average farm size went from 200 acres in 1950 to 400 acres. Census Bureau figures for 1959 show that 1.6 million farms grossed less than $2,500 that year, including wages from off-farm work and the value received from farm-produced food and housing. Note in the accompanying map the heavy concentration of low-income farms in the Southeast.

Three percent of the nation's farms accounted for half the total sales of farm production in 1975. This fact alone would have convinced Jefferson that "the laws of property" are in need of reform.

Ken first became acquainted with homesteading, or small-scale intensive farming, in the mid-1940s while stationed in the Philippines with the Army Air Corps. Chinese truck

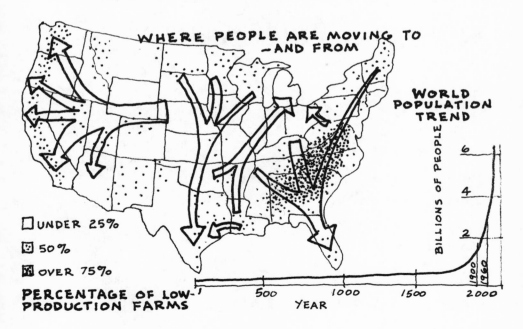

WHERE PEOPLE ARE MOVING TO —AND FROM

WORLD POPULATION TREND

BILLIONS OF PEOPLE

□ UNDER 25%

▣ 50%

▩ OVER 75%

PERCENTAGE OF LOW-PRODUCTION FARMS

YEAR

gardeners there practiced polyculture, or diversified farming. At the Agricultural College of Los Baños, Manila, Ken also became acquainted with the emerging Smahan Ng Masaganany Kakanin, or Samaka movement. (Roughly translated, the name means "united group effort to grow more plentiful food for families.") This was a program to reestablish war-displaced Filipino families in decentralized homestead communities using diversified farming techniques. Over 100,000 copies of the *Samaka Guide to Homesite Farming* have been printed in English since 1950, but it is now out of print, and the program is defunct under present government restrictions.

Another primary purpose of Samaka was to reeducate the

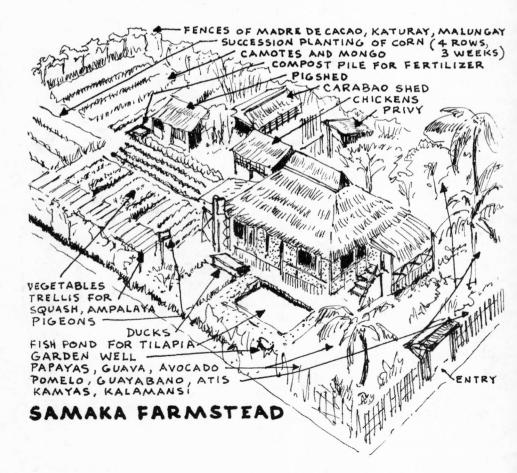

FENCES OF MADRE DE CACAO, KATURAY, MALUNGAY
SUCCESSION PLANTING OF CORN (4 ROWS, 3 WEEKS)
CAMOTES AND MONGO
COMPOST PILE FOR FERTILIZER
PIGSHED
CARABAO SHED
CHICKENS
PRIVY

VEGETABLES
TRELLIS FOR
SQUASH, AMPALAYA
PIGEONS
DUCKS
FISH POND FOR TILAPIA
GARDEN WELL
PAPAYAS, GUAVA, AVOCADO
POMELO, GUAYABANO, ATIS
KAMYAS, KALAMANSI
ENTRY

SAMAKA FARMSTEAD

Filipino rice farmer, persuading him to exchange large-scale farming for small-scale polyculture. The purposes of this movement were, of course, in conflict with the Rockefeller-financed International Rice Research Institute, which was busy developing strains of high-yield seed, which required extensive farming practice to produce.

In such instances we see the dichotomy that Third World countries face when accepting agricultural aid from American foundations and governmental agencies. This so-called aid has been little short of an economic tragedy for the rural poor in these countries. Such aid has become a windfall only for the already wealthy land-controlling monopolists of these regions. "Miracle rice" and "miracle wheat" have become a travesty for those who cannot afford the miracle. Who among the landless and hungry owns the tracts of land required for such specialized crops? Who possesses the advanced technological equipment for such production? Who can secure the funds for insecticide and herbicide programs that go hand in hand with this kind of farm management? And who can develop the vast irrigation systems heretofore unheard of in the Third World?

Upon Ken's return to the United States, he made a systematic effort to become familiar with all of the homestead literature and the emergent U.S. homestead movement. Through much of the 1950s we traveled separately to various homestead communities here and abroad, establishing lasting friendships with the movement's leaders.

When we became acquainted and started our homestead development in 1960 we were certain that the philosophy for this way of life was eloquently expressed by many but the techniques for achieving basic homestead goals were sadly lacking. Without exception, the how-to literature of the period was based on traditional farming practice, scaled down for the small operator. We recalled the efficiency and effectiveness of the Chinese truck gardeners and the Samaka movement and realized we would find no comparable Samaka-like agricultural assistance in the United States. Everyone in the American food industry seemed to have sold out either to petrochemical interests or to agribusiness itself.

Actually, there has been only one period in American history when a public agency was established specifically to aid the small farmer or the subsistence homesteader. That was during the Great Depression when the Roosevelt administration created the Farm Security Administration (FSA). Quite early in its operation the FSA challenged the cheap-labor plantation system of southern landlord farm interests when it attempted to deal comprehensively with rural poverty. It taught destitute farmers self-help farming and cooperative marketing techniques. During this time thirteen cooperative farmstead communities were established in the United States. These cooperatives were viewed as a threat to the concepts of free enterprise and private ownership of land, accepted hallmarks of America's mission and purpose. The co-ops failed.

The demise of FSA was brought about by the growth of the U.S. conglomerate agricultural administration. We use the term *conglomerate* to express the interrelatedness of the parts of the huge U.S. farm structure, from the county agent to the secretary of agriculture. The U.S. Agricultural Extension Service is but one segment of this public bureaucracy that actually serves the powerful, private American Farm Bureau Federation (AFBF). Forceful bureaucracies, along with agricultural land-grant colleges, were originally promoted by the U.S. Department of Agriculture. The whole system, financed by the American taxpayer, is influenced and used by agribusiness.

Corporate absentee landlordism, in the mantle of agribusiness, is the true beneficiary of the tax-funded agricultural research done by federal and state agencies and by agricultural land-grant colleges, using tax advantages originally legislated to aid and encourage individual farmers. In 1969 the Agribusiness Accountability Project calculated that, of the 6,000 scientific man-years expended by all state agricultural experiment stations, only 189 man-years are devoted specifically to "people-oriented" research. In other words, only 5 percent of the research effort of these tax-supported stations has been allocated to benefit the lives and work of rural people. The remaining 95 percent of this public-sponsored research benefits agribusiness.

The small landholder can expect little assistance from the county agent, land-grant colleges, or agricultural experiment stations, all of which are immersed in the study of plant and animal disease; insecticide, fungicide, and herbicide control; and the research and development of machinery for crop-picking. AFBF people are busy lobbying in state and federal capitals on behalf of big-farm interests against the right to strike by destitute, minimum-waged migrant workers. Administrators of the bureaucratic Department of Agriculture direct policies and disperse funds in support of ever more powerful agribusiness. Given a choice, the department favors bigness. For example, power reserves from the federally subsidized Tennessee Valley Authority (TVA) project can produce either nitrate or phosphate fertilizers. Nitrates are traditionally the small farmer's fertilizer, but large-scale farmers benefit more from the use of phosphate fertilizer. TVA produces only phosphates.

Throughout the 1960s we thought about the unfortunate status of the homestead movement. At that time, however, much of our energy went into raising children, into experi-

The Nelson homestead (Barbara's grandparents), Sweedhome, Nebraska, 1889

mental home-building, and into the compilation of Ken's first book, *The Owner-Built Home*. Since 1970, however, a rebirth of the homestead movement has taken place, exceeding the wildest expectations of observers. People are being driven to the country by the deteriorating conditions of life in the cities and suburbs. The world's chaotic political, economic, and social conditions stimulate many to seek more meaningful values in the relative stability and serenity of rural areas. Countless college students have become disillusioned with their profession- and consumer-oriented schooling. They are repelled by life-denying forces within the established system.

To meet the informational needs of those joining the rapidly growing exodus to the countryside, a number of peri-

The Kern homestead (Ken's grandparents), Glidden, Wisconsin,
1899

odicals and individual authors have begun to crank out
hackneyed, customary farming information. All the old
homesteading books have merely been republished in new
format. And as might be expected, failures among would-be
ruralists have become more common than successes. None of
these retread books or periodicals have proved of real help in
providing knowledge of organizational or manual skills. Nor
has there been the work-discipline to satisfy basic needs.

It is tragic to observe so many mentally and physically
qualified people failing in their attempts to live on the land. At
the outset they appear to have no concept of even the first
step—actual work—much less a concept of the whole complex of plant-animal-soil relationships or an understanding of

the techniques of the production, processing, and storage of food, which have challenged the most knowledgeable and experienced farmer. In the words of Dwight Eisenhower, the fate of this generation appears to have been sealed in its grandparents' generation: "I often think today of what a difference it would make if children believed they were contributing to a family's survival and happiness. In the transformation from a rural to an urban society, children are robbed of the opportunity to do genuinely responsible work."

Eventually, failure disillusions these new homesteaders, and they tend to react against the use of methods, materials, tools, and skills associated with living on the land and talk about isolated tribal communes, primitive living, and the like. These people despise mechanized, computerized urban existence, yet they are ill equipped to create a self-sustaining life on the land for themselves. So they assume that the next step should be isolation with fellow failures.

Thus, too, is born the esoteric notion of homesteading as a form of spiritual enlightenment. If one cannot grow plants, perhaps one can talk or listen to them. Dousing for water may

Anton Nelson (Barbara's father), Richvale, California, 1920

Hans Kern (Ken's father), San Diego, California, 1927

be simpler than understanding the complexities of the geologic structure of one's land. Planting by the moon dispels the need for a great deal of study about agronomy. Today, it appears that a book on homesteading must, to sell well, include some astrology or TM.

As we said at the outset, this book is different. It is an attempt to bridge the gap between conventional farming practice and a wholesome use of science and technology. After answering the whys and wherefores of homesteading, we intend to analyze the component parts of a balanced homestead environment, from human and animal shelter forms to crop production and maintenance functions. We also propose a descriptive evaluation of sensible techniques and routines for productive homesteading. Basically, we intend to encourage the postmodern way of country living, a life of self-reliance and at least partial economic self-sufficiency.

The greatest fine art of the future will be the making of a comfortable living from a small piece of land. *Abraham Lincoln*

1

Development Goals

Ignorance of fundamentals is a greater obstacle to progress than lack of popular appreciation of known facts.   *Whitman J. Jordan*

Look after the causes of things; the effects will take care of themselves.
        Eric Gill

Reduced to its simplest terms, a homestead is an ecosystem in which humans evolve in mutual association and coexistence with plants, animals, and other life forces. In this cohabitation the various components of the homestead germinate, develop, and mature at varying rates for varying purposes, all interdependent and individually supportive of life therein.

In the wider world of nature the evolving earth sends up volcanic ash and subsequently reduces parent rock to young soil, which matures, producing an environment for plant growth, and in turn, animal life. Through the process of natural selection, plants and animals adapt to their environment, in which, if allowed to make full use of nutrients, light, and water, they become resistant to extremes and flourish. In this manner vegetation achieves its climax growth, a process by which successive varieties of a species appear until the ultimate, environmentally suited form of the species appears. Hardwood and redwood forests and prairie grass are examples of the end result of this process. No "hunger signs" are evident in these stands, since even scarce nutritional elements are recycled from soil to roots, to leaves, and finally back to the soil. All the essential nutrients are utilized and conserved. This is the goal of nature and should therefore be the goal, and the criterion of success, of the mature homestead, where living in harmony with the environment also becomes a measure of one's personal maturity.

It may be assumed that people know what it takes to create a mature human life—one providing for the basic needs of the individual. Determining one's goals and proceeding with the means for achieving them requires varying degrees of mental perception and physical investment, control, and flexibility. People constantly perceive new needs and, to satisfy them, create new goals, utilizing new means for their realization. The process of human development is, at its best, dynamic and cyclic: the elimination of old, unsatisfactory results and patterns through experience and reevaluation, research, and design creates new, more satisfying results and patterns. This is the same process that gives rhythm, variety, and continuity to homestead development. Like a human life well lived, the creative homestead is truly a work of art.

Art is said to be the making of things for the right reasons. Homesteaders are artists to the extent that they make a thing for a purpose, for a right reason, and to the extent that they become fully involved in its creation. The thing created may be an entire homestead or just one element of that homestead. It might be a program for healthy soil or for plant or animal management, a tool promoting more efficient work, a walkway for safe, serviceable circulation between chore areas, or an intimate aesthetic space for human repose. The opportunities on the homestead for creative choices and artistic expression are multiple and intriguing.

One's whole person—body, mind, imagination, and powers of choice—is involved in organizing a homestead. To accomplish an artful order, the prospective homesteader must determine the purpose of the thing to be made. Then he or she must form an image and select materials to create the thing imagined. Finally, the homesteader must choose the right tool to shape the materials into the form desired. The ultimate result will be a fully functioning and self-sustaining homestead or part thereof.

Commenting on the human propensity for misdirection, English writer and sculptor Eric Gill called modern life as we presently choose to live it a "subhuman condition of intellectual irresponsibility." He justifiably asks:

How can it be agreed that food, clothing and shelter shall be produced *en masse,* by machinery, and simply as objects of merchandise, things produced solely for the profit of investors of capital, and yet that, fed on machine-made food, dressed in machine-made clothes, housed in machine-made buildings, we shall be able in our leisure hours—the hours when we are not working in the factories—to produce and enjoy the products of human civilization?

You cannot have responsible human beings in their leisure time if they are not responsible in their working time.

For working is the means to living, and it is life for which we have responsibility.

The focus is on combined thought and work. This dual approach to one's homestead achievement has more to do with success or failure than any other single factor. Only through thought and work does one find solutions to the probing questions of what to do and how to do it.

Before any amount of planning, thought, and work can take place, however, it is necessary to know exactly with what you have to work. A general survey and a penetrating assessment of your total resources is in order. Climate, soil condition, topography, vegetation, and animal stock are as important to homestead development as are the homesteader's personal abilities and needs, likes and dislikes. Then, after translating all of these elements into understanding, decision, and action, you must systematically, often daily, observe these components, as encouraged by M. G. Kains, author of many exceptional books on homesteading:

> One of the most profitable habits you can form is systematically, every day, to go over at least part of your premises in a leisurely, scrutinizingly thoughtful way and the whole of it at least once each week throughout the year to reap the harvest of a quiet eye and to fill the granary of your mind with knowledge of the habits of helpful and harmful animals, birds and insects; to observe and understand the characteristics of plant growth from the sprouting of the seed through all the stages of stem, leaf, flower, fruit and seed development; to note and interpret the behavior of plants, poultry and animals under varying conditions of heat and cold, sunshine and shade, drought and wetness, fair weather and foul, rich and poor feeding.

Relative to living a homestead life, two fundamental questions must be probed and answered at the outset: what will the land produce and what will it return to those who husband it? The answers will determine the degree of subsistence living that the homesteader can logically expect. Subsistence may be classified as *pure subsistence,* from which all essentials can be earned from the homestead environs; *quasi subsistence,* in which one-quarter of one's working time must be devoted to outside money-earning; and *semisubsistence,* in which half of one's earnings are derived from outside sources. The degree to which subsistence may be drawn from the land has a lot to do with family needs, desires, and life-style, and the homestead family must therefore anticipate and realistically assess the level of subsistence it requires.

Next to income, food is the largest item necessary to the homestead. Based on family need and influenced by family taste, it is essential that a family ascertain which plant and animal resources and how much of them will be used for its diet. Plants for family food consumption require specific garden and orchard plots, the development and management of water, soil enrichment, processing, and storage. Animal products require field crops and shelter for animals and the processing and storage of their products. Specialty crops, like fish culture, require provision as well, and the degree of mechanization by which these products will be produced must be determined.

Classifying homestead land is a priority for any survey. In order to make the best use of the land's productive capacity, it is necessary to make a map of the land, estimating its potential and calculating its limitations. In most regions of the United States, the Soil Conservation Service will prepare a land classification map free of charge. Whenever possible, however, homesteaders should prepare their own inventories, assembling known or observable facts about their land for their own map preparation. (Map-making is described in Chapter Three.)

Self-appraisal based on past experience, personal characteristics, and proficiencies is needed. Evaluating motivation, character, physique, and physical and emotional stamina, such an assessment should help to determine probable dropouts. The fact that homesteading is most often a family affair complicates any appraisal, but the lack of an essential characteristic in one participant may be made up by its occurrence in other members.

A yardstick for success on the farmstead may be indicated in responses to a questionnaire on what makes a good farmer first published in 1929 by the Bureau of Agricultural Economics. It is interesting to compare this early study with one made more recently.

1929 Questionnaire Response	1952 Questionnaire Response
1. Farm experience	1. Takes pride in farm and work
2. Wife's cooperation	
3. Ambition to succeed	2. Ambitious
4. Liking for farm work	3. Good manager
5. Getting work done on time	4. Plans his work
	5. On time with work
6. Hard work	6. Financially successful
7. County agent help	7. Builds up soil
8. Production management	8. Progressive
9. Farm papers	9. Good business judgment
10. Father having been a good farmer	10. Enjoys working with livestock

Today's homesteaders, often tumbling from the cracks and margins of the city jungle, are puzzling, indeed, to the landed farmer who inherits a long history of working the soil. Typical of the judgment leveled on these newcomers is the indignant comment of one farmer who was heard to declare, "Why, neither whiskers, nor weeds, nor uncastrated pigs annoy them!"

At this point goal-setting, surveying, and planning should be a hard-nosed effort to bring needs, desires, and personal and natural resources into line. This is the time to be realistic, for it is here that first failures occur. Too often the home-steader's limited resources are engaged in pursuit of un-realistic ends. Or impatient actions disrupt the balance and rhythm of steady growth. At any moment optimum develop-ment and ultimate goals may seem far from realization, but goals, too, grow on a homestead. Only a flexible approach will do; you must be grateful for unexpected results and patient with unintended variations. It is necessary to take a long view of homestead development, to imagine yourself creating a new plan for life on a plot of earth. In the words of Scott and Helen Nearing, "Each moment . . . should be treated as an occasion, another opportunity to live as well as possible. With body in health, emotions in balance, mind in tune and vision fixed on a better life and a better world, life . . . is already better."

2
Development Means

The most beautiful motion is that which accomplishes the greatest result with the least amount of effort.   *Plato*

Writers of homestead books are prone to warn that if a certain aspect of their work is not properly taken into account, one's entire homestead effort will likely fail. This admonition is neither false nor overworked. If it were not for the fact that our acquaintance has been more with failing homesteads than with successful ones, this book would never have been written. Site, water, soil, shop and tools, transport equipment, and balanced crops must all be considered for their individual and collective contribution to the success or failure of the homestead venture.

However, if we could predict a success-failure ratio for current homestead attempts, we would expect to find that disorder in one of the above aspects accounts for less than half of the fruitless endeavors, the majority of the bankruptcies being attributable to a twofold cause—the homesteader's ignorance or disregard of the physical and psychological means of development. You can somehow manage with a minimal water supply or battered equipment, but when physical stamina or poor self-image fails to command clear direction, concerted effort, and routine performance, you may just as well forgo the endeavor at its outset.

It is essential that you be aware of the need for physical strength and stamina. Similarly, you need to allocate your *time* realistically, to have a comprehensive resolute *plan,* and to be highly motivated in your *work*. Systematically searching for the simplest, most direct way of doing the work at hand assures that you will accomplish your goal. An example of this is using deep littering in an animal shelter to eliminate the chore of frequent barn-cleaning. Or draining oil from the crankcase of a vehicle when it is hot, just after it has been driven a while, will empty it quickly and more thoroughly.

Homestead tasks are easier, simpler, faster, and more enjoyable when buildings and equipment are conveniently laid out and when work movements are simplified. The premier study for the effective use of labor, today called time-motion study, was made by F. M. Taylor in 1898. An eastern steel company hired Taylor to find ways to increase pig iron production. He began by examining the process of shoveling raw material by hand. He found that when shovel capacity was reduced from 30 to 21 pounds, production rose from 12 to 47 tons a day. Taylor's only other suggestion to the company was for management to observe workers' rest periods. Stamina depends partly on the rhythmic functioning of the human biological time clock, which reaches a peak at 9 A.M. and a low at 3 P.M.

It is surprising how few people are aware of the simple body mechanics used in the act of shoveling, a movement employed almost every day on a homestead. Working habits are formed at an early age, becoming automatic and difficult to change in later life. Energetic youth consequently shovels its heart out, wasting time and energy and forming lifelong improper work techniques. Later in life, when it becomes essential to conserve energy, work habits instilled early in life fatigue the body and may even be injurious to it. Therefore, some important shoveling concepts follow.

When you are shoveling, your back should remain straight. When you are reaching down to load a shovel, bend ankles,

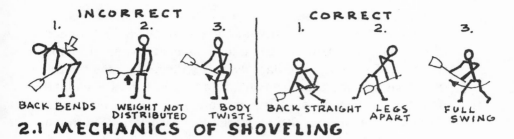

INCORRECT   CORRECT

1.   2.   3.   1.   2.   3.

BACK BENDS   WEIGHT NOT   BODY   BACK STRAIGHT   LEGS   FULL
              DISTRIBUTED   TWISTS   APART   SWING

2.1 MECHANICS OF SHOVELING

knees, and hips. Spread your legs somewhat apart, one in front of the other, to balance the weight on the shovel. As you straighten, transfer your weight to the back leg. As the shovel load is thrust forward, transfer weight from the back leg to the front leg. This provides for a full, even swing not involving the weighted momentum of the body for its full execution.

Once the mechanics of shoveling—or of any task, for that matter—are learned, you will develop a rhythm so that your performance becomes smooth and automatic. Any forward-backward movement is desirable when working. In this manner contrasting groups of muscles (flexors and extensors) work and rest alternately. It is wise to work with minimum movement, never using the whole body when a part can accomplish the same purpose. When each part of the body is being used, the energy expended is proportionate to the task to be done. Therefore, employ only the set of motions that will accomplish the work at hand. Sequential movement begins with the movement of the finger, moving to the wrist, forearm, upper arm, and trunk.

At the same time you use an economy of means, your whole body must be integrated into task performance. Correct breathing, for example, must accompany every movement. Improper breathing routinely accompanies lazy, slumping posture, and poor posture affects use-of-self.

In the past *posture* had a very static connotation, implying only one's general bearing—how one sat and stood. British author F. Matthias Alexander *(The Use of the Self)* widened the concept of posture to include all bodily movement, using the term *use-of-self* to replace *posture*. Alexander rejected the idea that any one exercise could improve posture. Instead, he felt that breathing supports movement just as movement reinforces breathing, sitting, standing, or working. You must master integrated body movement in order to become an effective worker.

During childhood, when we are beseeched to tuck in our chins, thrust out our chests, and stand up straight, we are paradoxically introduced to the archaic custom of chair sitting, which we proceed to do the rest of our lives when not standing, moving, or reclining. Alexander has called chair sitting "the most atrocious institution, hygienically, of civilized life." The consequence of this indoctrination, he claims, along with improper diet and poor self-concepts, has accentuated the forward tipping of the base of the spine, the collapse of the chest upon the trunk, and the drooping of head and shoulders. This early conditioning has produced a populace with swayback, rounded shoulders, and hanging head.

Physiologically speaking, the small of the back, like the keystone of an arch, is the primary support for our heavy trunk. When the body is employed in work, the entire spinal column is lengthened; that is, it is stretched into a curve by the strong muscles of the back and trunk, just as it is when we assume a squatting position. Movement in work of any kind should involve the greatest lengthening of the spine, not its arching. But our pampered, civilized habit of chair sitting does just that. Its constant arching of the spine pulls thoracic vertebrae in, constricting lung capacity and affecting the rate and quality of our breathing. Stomach, liver, and other vital organs are, consequently, subjected to unnatural compression. Lumbar vertebrae at the small of the back are sucked in, and sacral vertebrae must transmit forces that are applied transversely. Sexually the result is disastrous. In males there is a predisposition to hernia, which is the extrusion of viscera through weakened abdominal walls. In females there is a propensity for a slumping, tipped uterus, making childbirth and female functioning difficult. Intercourse is hardly enhanced.

The figures of the working couple in Hagborg's *October Potatoes* (Figure 2.2) illustrate the deprivation of poor posture. This painting was done on the eve of the Industrial Revolution and indicates to some extent why machine technology was welcomed by those to whom hand labor was so burdensome; these were hard times in western Europe. The painting shows none of the upper body's outreaching expansiveness—so essential to dynamic joy in living. The couple seems concerned about a full potato sack, not about quality in life. Today we can have both.

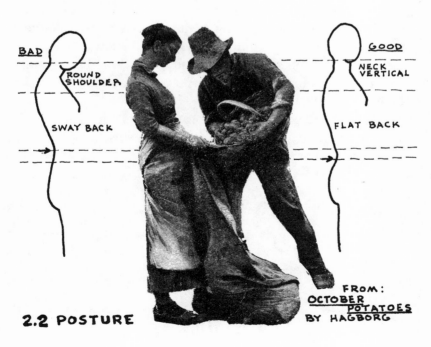

BAD

ROUND SHOULDER

SWAY BACK

GOOD

NECK VERTICAL

FLAT BACK

FROM: OCTOBER POTATOES BY HAGBORG

2.2 POSTURE

When picking up heavy objects about the homestead, your back should be flattened, your spine stretched. When bending your knees to grasp the object, your back will remain straight if your neck is stretched and your chin pulled in. Then, while bending forward from hip joints, you can lift the load vertically with the simultaneous straightening of knees, hips, and back. Alexander's technique distributes a work load over as many joints and muscles as possible, putting no more effort into the work than required. If, when picking up a 100-pound load, you stoop without bending your knees, the weight exerted on muscles and vertebrae of the lower back is on the order of 1,500 pounds!

Little effort has been made to redesign homestead implements and equipment to promote use-of-self. One exception, however, is a new bucket design developed in Britain. About one-third of the vertical capacity of this container was lopped off, altering its center of gravity. The flattened side could then be carried closer to the body, reducing carrier fatigue. We also favor a wheelbarrow design that places the center of gravity well forward, so that the load is not carried by the handles. More widely spaced handles, along with a lower center of gravity, provide better balance as well. You can readily convert commercially made back-buster wheelbarrows in the homestead workshop.

Well-designed equipment and implements may be rare, but there is much that you can do to reduce work fatigue. Con-

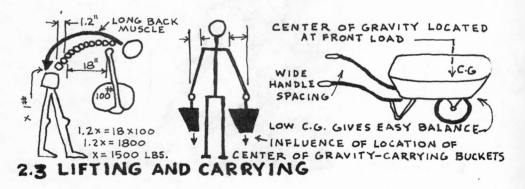

2.3 LIFTING AND CARRYING

venient building layout and chore routes are discussed in Chapter Four. It should be mentioned here, however, that planning work ahead of time will reduce unnecessary trips. Feed supplies should be dispensed by gravity whenever possible. Movement should be made in a continuous curving motion rather than in a straight line, involving sharp changes of direction.

Body performance is enhanced by a conscious attempt to reduce effort and travel. For example, you may throw an ear of corn to a pig with a full-arm movement, but it would be better to use a simpler, shorter movement of your wrist and forearm. Arrange short reaches for accuracy, speed, and endurance. Keep frequently used equipment nearby. A study was once made of picking fruit by hand. It was found that a simple shoulder-supported carrying device that allowed the picker's hand to be free boosted production 35 percent.

Efficient production requires two preliminary decisions: what to do and how to do it. Careful scrutiny and analysis will indicate to the homesteader whether or not a given operation should be maintained or relinquished. If the operation is indispensable, perhaps it can be combined with other operations. If not, perhaps its sequence can be reordered and simplified.

Handwork, such as that involved in food preparation or food-processing, requires specially planned work centers. At each level you must measure for the amount of space required for equipment and supplies. Heights should be relative to your size and ability to utilize those heights at each corresponding level. Then arrange equipment and supplies in the sequence of their use for a specified task.

Human behavior is the muscular or glandular response to perceptual and sensory stimuli; it is one's conduct. Essentially, behavior is movement, both physical and emotional

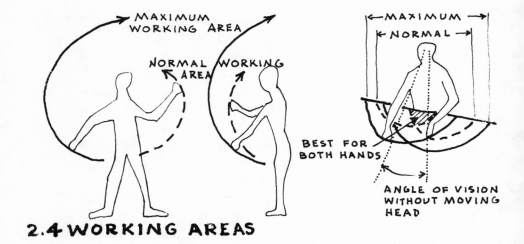

2.4 WORKING AREAS

(*emotion* means "movement outward"). Any disruption of emotional current restricts that movement, blocking the flow of energy to vital organs, especially to the brain and genitals.

Perceptual and physical behavior forms the inseparable duality, the integral whole of a person. It is, however, self-perceptions—which begin with birth—that result in our physiological structuring. We actually create around ourselves a protective structure, an armoring of guarded emotional reactions and rigidly controlled physical responses. Throughout life we project this image outward, convinced ourselves that this is "the real me."

To nurture those characteristics necessary to a healthy homesteader, we must become familiar with our most cherished, suppressed attitudes and perceptions and our body's resulting response, its "language." When performing tasks, we must become aware of how we are holding our body, stretching those parts that are tightly held, breathing into them, and integrating them with our inner and outer awareness experience. When coupled with economy of movement, this awareness frees static posture and stiffened muscles. Work demands and tensions are engaged and resolved when they occur. As perceptions improve, posture improves. Breathing, hearing, and vision are also aided. We can then master the skills associated with homestead work activity, luxuriating in the satisfaction of a job well done.

Freeing static feelings inspires the personal renaissance of imagination and creativity. Just as standing tall makes room for the expansion of inner air, thinking tall makes room for the expansion of view of self and view of the world. The homestead family as a whole may then develop a harmony among its members and with all elements of the living environment. Only then will the homestead flourish.

3

Selecting the Site

Buy land. They ain't making any more of that stuff.　　　　*Will Rogers*

Anyone who has recently shopped for land knows of the increased demand for rural property. The demand rose 40 percent on a national average between 1970 and 1975. In general, states with the highest population density have the most expensive land. Other conditions being equal, it would be better to choose land costing $50 per acre in a state with a density of five people per square mile than land costing $1,500 per acre in a state where density is 1,000 people per square mile.

Figure 3.1 has been compiled from the 1970 U.S. census and illustrates the close relationship between population density and rural land value. Between 1965 and 1970 the percentage of increase in land value got higher as its price got higher. During the Great Depression a similar demand for land had its origins in unemployment and insecurity. The origins of today's exodus to the countryside are more sophisticated. Industry is dispersing from its urban centers, and suburbia and urbanity are moving outward. An estimated 1 million acres of rural land are engulfed yearly by residential, industrial, highway, and other nonfarm use. Commercial farms are expanding further in an attempt to make the investment in large-scale machinery more efficient and economical.

The amount of land now being withdrawn from the market

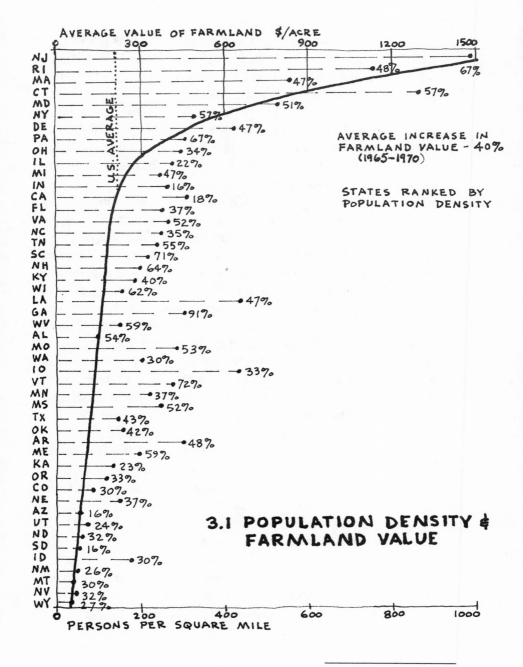

AVERAGE VALUE OF FARMLAND $/ACRE

AVERAGE INCREASE IN
FARMLAND VALUE - 40%
(1965-1970)

STATES RANKED BY
POPULATION DENSITY

3.1 POPULATION DENSITY &
FARMLAND VALUE

PERSONS PER SQUARE MILE

for speculation is significant, too. Experience has shown that land investment is particularly high during and immediately after a war, for land is then considered a safe hedge against inflation. Today, following the cessation of war in Vietnam, the purchase of land that is desirable to speculators is not practical for the prospective one-horsepower homesteader for reasons that will soon become evident. It behooves us to locate areas of land adequate for homesteading that are not attractive to investor speculation.

Brief attention to appraisals of rural land show clearly the various property features that are of speculative value. Proximity to rail transport or main roads may be important to the commercial farmer but is probably not worth the additional cost to the small-acreage homesteader. Level ground is far more valuable for land speculation than sloping ground, for—as any speculator selling to large-scale farming interests knows—level ground is necessary for uniform crop cultivation, making possible the use of power equipment. But small-acreage homesteaders find that, for their purposes, drainage tends to be poor on level ground and that accumulated water may leach nutrients down to deeper layers, leaving hardpan in its wake. This crust, resulting from the poor aeration and stripping of soil, becomes impervious to plant roots.

Hilly locations are unsuitable for large-scale farming operations but advantageous to the potential homesteader. During daytime hours a level, protected valley receives reflected sunlight from surrounding slopes and is, in hot weather, considerably warmer than surrounding hillside areas. At the same time, wind movement over hilly regions provides better ventilation and results in less heat buildup. At night, air drainage down slopes is accelerated and temperature inversion takes place. In cold weather reservoirs of cold air drain from surrounding slopes into low-lying basins.

Some years ago in Ohio climate comparisons were made of a valley and an adjacent hill area. The hill site, which consisted of a grottolike impression weathered from the face of a cliff, was located at the cove end of the valley. The hill site enjoyed 276 frost-free days, while below in the valley frost pocket there were only 124 frost-free days. Maximum-minimum temperatures for the hill site were 75 and 14 degrees Fahrenheit, respectively, while the valley was subjected to a range of 93 to 25 degrees.

In the northern hemisphere slopes facing south receive more insolation (solar radiation) than northern exposures. The degree of slope determines the amount of radiation received. During the growing season the warmest slope is the one most nearly perpendicular to the sun's incoming rays. Steepness of the site to be chosen should, therefore, increase with a move to a more northerly latitude. According to the U.S. Department of Agriculture, land in southern Idaho with a 5-degree southerly slope has the same solar climate as level land 300 miles to the south in Utah. Conversely, ground with a 1-degree northerly slope lies in the same solar climate as level land 70 miles north of the Idaho location.

Figure 3.2 shows how light beams of similar size striking a northern slope and flat open ground cover larger areas than those striking a southern slope. The larger the area covered, the less heat per unit is received. The accompanying map indicates that the western half of the United States has more prominent physical features than the eastern half of the country. Generally, the climate in the West alters with changes in altitude. In the East weather changes depend more upon north-south latitude.

A southwesterly slope is warmer than a southeasterly slope. Sunshine on a southeasterly slope is slower to affect warmth after prolonged night cooling, partly because evaporation of

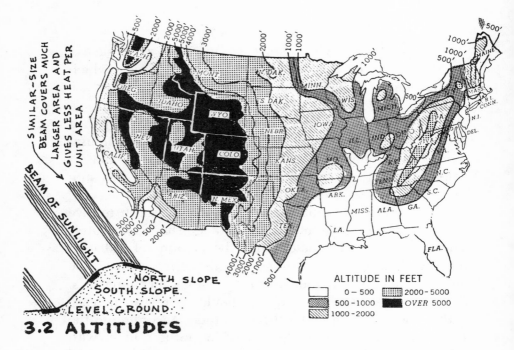

3.2 ALTITUDES

ALTITUDE IN FEET

0 – 500
500 – 1000
1000 – 2000
2000 – 5000
OVER 5000

morning dew uses some of the sunlight's energy. The westerly slope of a hillside has higher average air and soil temperature and a longer frost-free season. Injury to plants from extremes of cold is greater on slopes facing east, while heat is more injurious to plants on slopes facing west. Those facing north tend to be more moist than comparable ones facing south.

As mentioned previously, climate considerations are different for the eastern and western regions of the United States. In the East frost differentials are determined by north-south latitude, whereas elevation determines temperature differentials in the western half of the country. Latitude in the East and elevation in the West also determine the length of the growing season. (Growing season is the average period between the last killing frost in spring and the first killing frost in fall.) The limited growing season, for example, would

be the primary restriction when choosing a site in Alaska. Climate extremes (excessive heat or cold and wide variations in
the amount of sunlight or rainfall) and their duration must be
considered when choosing a homestead site.

Along with climate and topography, soil type is of foremost
importance to homestead location. Soil classification is an involved subject and will not be discussed here. But a few
general pointers about soil should assist your selection of a
site. Dark soil usually indicates high fertility. Gray or yellow
soil connotes poor drainage as well as low fertility. You should
look for medium-textured soils. You may expect that soils
high in sand or clay will be relatively unproductive. In spring
sandy soil thaws before, and warms faster than, clay. This is
because of its higher heat capacity, its higher thermal con-
ductivity, and its reduced propensity for evaporative chilling.

You can determine soil fertility by observing plant growth.
Fast-growing weeds such as giant horseweed or cocklebur
denote good soil condition. Red sorrel grows in poor, acid soil.
If a plant has deep, rich color, the soil in which it grows is
probably quite fertile. Tree limbs that extend upward rather

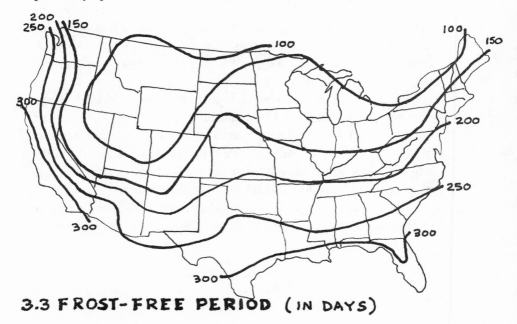

3.3 FROST-FREE PERIOD (IN DAYS)

than droop indicate fertility. Walnut, cypress, white oak, and cottonwood trees are indicative of good soil; blackjack and pine tend to grow in poor soil.

Seldom is a homestead site favorable in all respects. We therefore learn to compensate and adapt to shortcomings. If latitude fails to moderate temperature, perhaps resulting in occasional freezing, you may make up for this handicap by increasing the altitude of your site. A sloping site oriented to the sun may compensate for its more northerly latitude. It may even be necessary to supplement low annual rainfall by locating in an area enhanced by dew or fog or by living close to sources of irrigation or water storage. Good soil texture may substitute to some extent for inadequate moisture. You should select crops for their adaptability to your soil; asparagus, for example, thrives in sandy soil while heavy soil is too wet for it.

When selecting the actual homestead site, evaluate successively the region, state, county, and community until you arrive at a prospective site. The most important tools for this are maps, which can reveal much. Start with a set of topographic maps from the U.S. Geological Survey. They are accurate and show topographical features. There are maps for about half of the United States. Much desert and prairie land, however, has similar topography and climate, so maps covering such areas in their entirety are not made.

Maps are one of the best tools for site exploration. They are also sources of continual fascination, for nothing else, short of earth contact, offers as much understanding and appreciation of land. Studying them is one of the finer joys of looking for a site. Folding maps is, however, as difficult as cutting bread; maps should be rolled and bread should be broken.

The government prints detailed climate booklets for much of the country. These publications show monthly maximum and minimum temperatures, annual precipitation, length of freeze-free period, prevailing wind direction, and relative humidity. The Soil Conservation Service can supply you with aerial photographs and soil maps for many regions. The Bureau of Public Roads has informative county highway maps and the Postal Service has maps showing rural mail routes.

You can find information about a prospective acquisition on public record at county offices. The local title insurance company often has more up-to-date title information than these offices, but information about assessed valuation, the amount of taxes paid, special assessments for drainage, and the like can be found in county record books. Dates and amounts of sale for adjoining properties are also in these records. Individual property owners' names are indicated on county plat books. Addresses can be found in the assessor's office.

For information about remote properties it may be necessary to draw site information from government township plats. These are available from the U.S. General Land Office or are on file with the state auditor. Individual counties may have government township plats on file. Check with your county surveyor.

Once you narrow the land search to a specific county or community, you should take up residence there and begin the quest for a specific site. To start, inspect the tax rolls in the county treasurer's office for properties on which taxes have remained unpaid for a number of years. Unwanted property can often be bought for the amount of the unpaid taxes. The county tax collector also keeps on public file a list of parcels with unpaid taxes. A letter to one of these presumably "distressed" property owners will sometimes uncover some prime but inexpensive homestead property.

Banks and trust companies are engaged in liquidating property at bargain prices. Auctions are another good source, or you can advertise for property in the local newspaper. It is a good practice to become acquainted with people in the community of your choice. Ask for available land, and let it be known that you are in the market to buy. As a final, somewhat desperate resort, roll up your maps and visit your friendly real estate broker, remembering that these agents work for the seller, not the buyer.

Something should be added to this subject of unwanted or distressed property. People (and particularly professional land speculators) are unimaginative when it comes to developing a problem site. They are interested in a quick turnover for their money and not in excavating, filling, terracing, planting, or any other possibilities for adapting land to fit their needs. Homestead site requirements are flexible and adaptable, not static like those of tract-housing developments or commercial farming enterprises. It is good management to capitalize on the fact that you can profitably utilize a piece of land that nobody else wants.

When you have located the homestead site, you should start transfer proceedings by making an appraisal map. This map is drawn mostly from on-site inspection. It should show an outline of the property according to its complete legal description. Important topographic and natural physical features, such as streams, fields, and woods, should be indicated. Any improvements such as buildings, fences, or roads should be shown. A tentative homestead layout and land-use sketch can be suggested on this map.

A photo and a tentative map for a prospective land purchase are depicted in Figure 3.4. As it happens, someone actually purchased the site illustrated, and our Owner-Builder Design Group proceeded to plan a model homestead specifically for this site. We will make reference to this model homestead throughout the book. The inventory map is illustrated in Figure 4.3.

3.4 APPRAISAL MAP

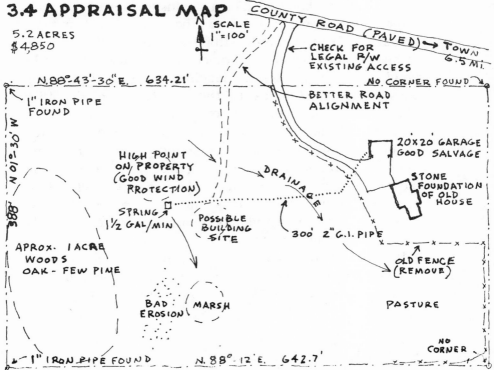

5.2 ACRES
$4,850

SCALE
1"=100'

COUNTY ROAD (PAVED) → TOWN
6.5 MI.

← CHECK FOR
LEGAL R/W
EXISTING ACCESS

No. CORNER FOUND

BETTER ROAD
ALIGNMENT

N.88°-43'-30"E. 634.21'

1" IRON PIPE
FOUND

N 01°-30' W

388'

HIGH POINT
ON PROPERTY
(GOOD WIND
PROTECTION)

DRAINAGE

20'×20' GARAGE
GOOD SALVAGE

STONE
FOUNDATION
OF OLD
HOUSE

SPRING
1½ GAL/MIN

POSSIBLE
BUILDING
SITE

300' 2" G.I. PIPE

OLD FENCE
(REMOVE)

APROX. 1 ACRE
WOODS
OAK - FEW PINE

BAD
EROSION

MARSH

PASTURE

1" IRON PIPE FOUND   N. 88°-12'E. 642.7'

NO
CORNER

At this point it would be prudent to confer with those county officials who may be involved in approving your land transfer. At the county recorder's office you can find out if the property can be legally transferred. Most states have land-division regulations, and many counties require a legal land survey before property may be sold. Then check with the county planning office for possible zoning restrictions. Find out, too, about building restrictions. The county health department may even have something to say about sanitation requirements in your area.

By this stage in the transaction, you will probably know your way around county offices on a first-name basis. You may, therefore, attempt the title search of your prospective acquisition yourself. Title insurance companies customarily perform this service—for a generous fee. They issue a mortgage policy that protects only the value of the land. If you build a $20,000 homestead on a site insured for $1,000 and a missing heir later arrives to cloud the title with a claim to the property, you recover only $1,000 from title insurance. The whole operation is costly and ridiculous, because with very little effort you can check the legitimacy of your title. Merely consult the tract index in the county recorder's office. Most counties keep an abstract of title on record. This is a condensed history of all recorded transactions for the parcel of land that you are buying. By examining the abstract, drawing up a simple deed, and preparing a closing statement (payment), you can keep several hundreds of your dollars from reaching the title officer, an escrow agent, or possibly a real estate lawyer, who generally charges 1 percent of the purchase price for services.

If the transaction is not made in cash, the simplest method of land transfer is that in which the seller supplies credit to the buyer. Either a deed is given to the buyer, with the seller

3.5 COMPARATIVE HOMESTEAD LAND POTENTIAL

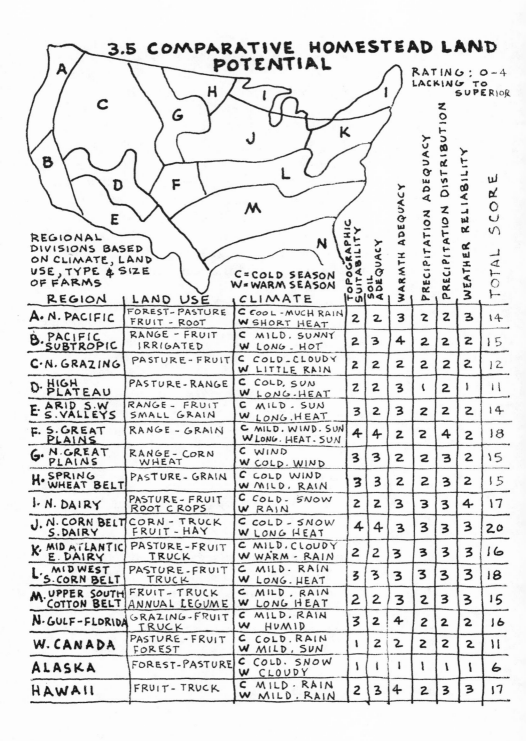

RATING: 0-4
LACKING TO SUPERIOR

REGIONAL DIVISIONS BASED ON CLIMATE, LAND USE, TYPE & SIZE OF FARMS

C = COLD SEASON
W = WARM SEASON

REGION	LAND USE	CLIMATE	TOPOGRAPHIC SUITABILITY	SOIL ADEQUACY	WARMTH ADEQUACY	PRECIPITATION ADEQUACY	PRECIPITATION DISTRIBUTION	WEATHER RELIABILITY	TOTAL SCORE
A. N. PACIFIC	FOREST-PASTURE FRUIT-ROOT	C COOL-MUCH RAIN W SHORT HEAT	2	2	3	2	2	3	14
B. PACIFIC SUBTROPIC	RANGE-FRUIT IRRIGATED	C MILD, SUNNY W LONG-HOT	2	3	4	2	2	2	15
C. N. GRAZING	PASTURE-FRUIT	C COLD-CLOUDY W LITTLE RAIN	2	2	2	2	2	2	12
D. HIGH PLATEAU	PASTURE-RANGE	C COLD, SUN W LONG-HEAT	2	2	3	1	2	1	11
E. ARID S.W S. VALLEYS	RANGE-FRUIT SMALL GRAIN	C MILD-SUN W LONG, HEAT	3	2	3	2	2	2	14
F. S. GREAT PLAINS	RANGE-GRAIN	C MILD, WIND, SUN W LONG, HEAT, SUN	4	4	2	2	4	2	18
G. N. GREAT PLAINS	RANGE-CORN WHEAT	C WIND W COLD, WIND	3	3	2	2	3	2	15
H. SPRING WHEAT BELT	PASTURE-GRAIN	C COLD WIND W MILD, RAIN	3	3	2	2	3	2	15
I. N. DAIRY	PASTURE-FRUIT ROOT CROPS	C COLD-SNOW W RAIN	2	2	3	3	3	4	17
J. N. CORN BELT S. DAIRY	CORN-TRUCK FRUIT-HAY	C COLD-SNOW W LONG HEAT	4	4	3	3	3	3	20
K. MID ATLANTIC E. DAIRY	PASTURE-FRUIT TRUCK	C MILD, CLOUDY W WARM-RAIN	2	2	3	3	3	3	16
L. MIDWEST S. CORN BELT	PASTURE-FRUIT TRUCK	C MILD, RAIN W LONG, HEAT	3	3	3	3	3	3	18
M. UPPER SOUTH COTTON BELT	FRUIT-TRUCK ANNUAL LEGUME	C MILD, RAIN W LONG HEAT	2	2	3	2	3	3	15
N. GULF-FLORIDA	GRAZING-FRUIT TRUCK	C MILD, RAIN W HUMID	3	2	4	2	2	2	16
W. CANADA	PASTURE-FRUIT FOREST	C COLD, RAIN W MILD, SUN	1	2	2	2	2	2	11
ALASKA	FOREST-PASTURE	C COLD, SNOW W CLOUDY	1	1	1	1	1	1	6
HAWAII	FRUIT-TRUCK	C MILD, RAIN W MILD, RAIN	2	3	4	2	3	3	17

taking back a mortgage; or the sale is made under an install-ment purchase contract. In the latter case the legal title remains with the seller until all or a specified portion of the purchase price has been paid. This contract of sale is pre-ferred by the seller over a mortgage contract. In cases of default a mortgage contract requires an expensive foreclosure sale; a contract of sale is merely terminated.

Most stationery stores carry deed-of-conveyance forms. After the deed is made out, sign it before a notary public and record it in the county recorder's office. Then, when you move onto the land, file a homestead exemption with the county recorder. Most state legislatures have adopted this statute to protect the value of the family home from creditor claims.

At some point in the procedure you will want to check or es-tablish homestead property corners. Again, with little exper-tise you can dispense with the service of yet another un-necessary professional. Land-surveying is expensive—and expendable.

The thirteen original colonies of this country used a metes-and-bounds survey. This is the simplest survey technique to retrace, since it starts from a known point and goes a set distance and a set bearing to the next point. *Metes* are merely measurements of length, and *bounds* are either natural or man-erected boundaries, such as streams or roads.

In 1785 the government adopted the rectangular survey. This technique is used in twenty-nine states (see Figure 3.6). With this kind of land division a north-south meridian line and an east-west base line are first established. The intersec-tion of the meridian and base lines forms the corner of four 24-mile squares, called townships. Each township is divided into 36 1-mile squares, called sections. Section corners and half-section corners were originally set by the U.S. Land Of-fice, now located in every state capital. Missing corners can often be found or reset by retracing the original survey notes. Land parcels can be surveyed out of sections by starting from known section corners and following the bearings and distances established in the original survey.

The only tools you need to survey your homestead are a 100-foot steel tape, a pocket compass, and maps or government notes. The compass should be the type that rotates with respect to the box in which it is mounted. As Figure 3.6 illustrates, the map maker can then turn a circle through an angle equal to the magnetic declination. The bearing you observe will thus be true and not magnetic. East of the line of zero declination the north end of the compass points west of north. West of that line it points east of north.

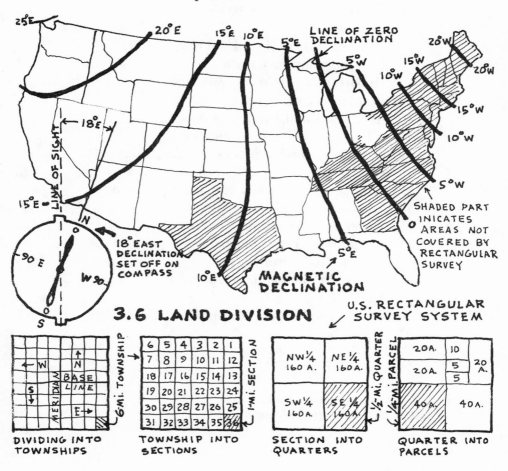

3.6 LAND DIVISION

Selecting a site that satisfies your homestead needs requires care. Many factors need thoughtful consideration. This chapter may fall short of mentioning all of the factors you may encounter in individual circumstances, but the following checklist provides a start for evaluating those items judged most important to appropriate selection of a homestead site.

Checklist for Selecting a Homestead Site (in order of importance to the authors)
1. Adequate domestic water supply
2. Effective solar exposure
3. Sufficient space
4. Satisfactory growing season
5. Pure air
6. Reasonable cost of land and taxes
7. Favorable natural terrain
8. Opportunities for local employment
9. Fertile soil conditions
10. Availability of natural resources
11. Adequate precipitation and drainage
12. Congenial neighbors and minimal neighborhood nuisances
13. Conducive zoning and building regulations
14. Reasonably stable cost of living
15. Natural beauty of area
16. Status of local and state politics
17. Potential for power supply
18. Access for roads and transportation
19. Facilities for health care
20. Cultural and educational opportunities
21. Recreation

4

The Homestead Plan

As, in nature, variety gives richness and interest to landscape, to sky, to the vegetable and animal kingdoms, so, in art, it adds to the interest of the whole by the diversity which it affords in the arrangement, sizes, or forms of the different parts.
    Andrew Jackson Downing, The Architecture of Country Houses

The key word in this quotation is, of course, *variety*—the concept of which, in its fullest meaning, forms the fundamental thesis of this book. As we elaborate in following chapters, some famous agriculturists base their premises for a healthy agriculture on variety or *polyculture*. Downing, a famed agriculturist himself, employed variety in architecture as well. In this chapter we propose an examination of this view as it relates to the conception of a homestead plan.

The entire homestead, from its structural design to its various management programs, is expressive of the homesteaders' view of life. The design of the simplest farm building and the programs for healthy plants and well-cared-for animals reveal the feelings and thoughts of their creators. The homestead design proposed herein is therefore based on a particular set of life principles deemed self-evident and essential to our way of thinking and living.

People often approach their homesteading project with a narrow concept of it rather than with a realization of its all-encompassing nature. With their primary impetus an escape from city life, they plan, at least, to garden "organically." This is not what homesteading is all about. Familiarity with organic gardening is insufficient preparation for this life. A *long view* of homesteading must precede its disciplined execution. Assuming life-affirming endeavor to be the frame of reference for a fully functioning, self-sustaining homestead environment, plants, animals, and humans may then proceed to live in mutual association there. The planned homestead therefore embraces and supports within its domain a population of interdependent beings whose dynamic response to their interaction forms the homestead ecosystem.

In this book we are thinking in terms of building a living system in which the entire homestead becomes a maturing organism in itself, rather than just a static plan for a complex of buildings and a physical layout—an imposition on the land, so to speak. The animal, plant, and human constituents of this system will grow and change, and so will their various needs as they move toward maturation, the hallmark of the living homestead. As your homestead grows and changes you will become aware that no other homestead on earth is precisely like it.

4.1 HOMESTEAD ECOSYSTEM

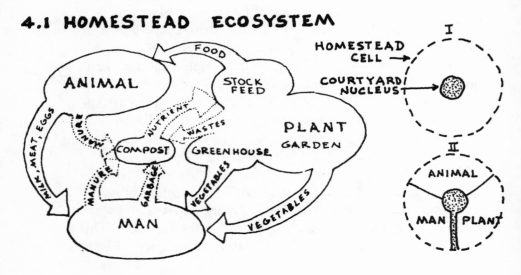

At the beginning homestead-planning requires a survey map of your property. First walk over the land, carefully examining all significant variations. Soil variances can be determined by examining a small sample of earth, procured by an auger. This sample will indicate topsoil depth, texture, permeability, moisture capacity, inherent fertility, organic matter content, and other characteristics that affect the management and productivity of soil. Simple visual observation will also indicate the slope of your land and the degree of its erosion, dampness, and drainage. You should record all of this information in notes or directly on the map.

Draw a plot plan of the property to a scale of 1 inch equals 50 feet or more. Lightly color the various areas as follows: forest and woodland, dark green; meadowland and permanent grasses, light green; arable land, brown; rough hill pasture, yellow; garden and orchard, purple; agriculturally unproductive areas, red; ponds, ditches, and streams, blue. Indicate north on the map. Show existing and proposed road access, water and electrical power sources, and any existing buildings, fencing, and other improvements. Draw arrows specifying winter wind, summer breeze, and view direction of your preferred homesite. You can use directional arrows to show water and drainage flow.

Additional facts that should appear at some place on the plot plan are latitude and longitude, elevation above sea level, maximum summer temperature and minimum winter temperature, and average number of growing days per year. A general description of soil conditions is useful when classified as to the amount of clay, silt, fine sand, coarse or fine gravel, and stones. You should note some indication of soil depth and drainage, as well as the amount of surface organic material. Also indicate areas that have a tendency to erode.

Possibly the most important design aid to be included on

this map is contour lines. Establish the known points all having the same level. Then, as seen in Figure 4.3, draw lines to connect these points. The lines designate the sculptured surface of the ground and should be drawn on the map at 10-foot intervals.

Two people can prepare a contour map using three basic tools: a 100-foot steel tape, a Brunton pocket transit that incorporates a compass, and a long wooden pole (called a *rod*) upon which you have marked inches and feet. The tape is used to measure horizontal distance. The transit is used to sight a level line from a position of known elevation and to take the bearings of that line. The rod helps you record increases or decreases in elevation from that line, thereby establishing the elevation of points surrounding the position of known elevation.

4.2 HOMESTEAD SURVEY

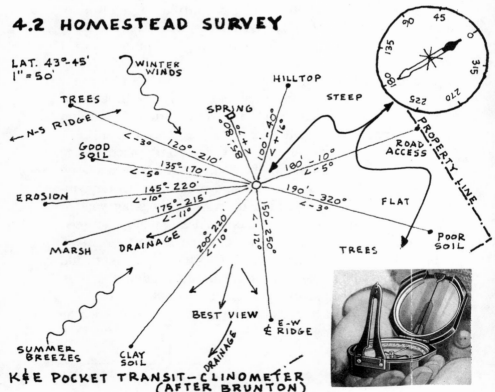

K&E POCKET TRANSIT—CLINOMETER
(AFTER BRUNTON)

Actually, the true elevation above sea level for this *position of known elevation* may not be known and, in fact, need not be known. The position serves simply as the starting point for the contour map and may very well be given an arbitrary value, such as 0 or 100. What is important is that you accurately locate or determine this point on the map, since all other points will be located relative to it, both in position and in elevation. To simplify the map-making process, it may be practical to use the highest point of the homestead for a starting place; then all subsequent elevation readings can be subtracted from the elevation of this position. However, this may not be practical in every situation.

The actual process of taking the readings is easier done than said, and we hope what follows, along with Figure 4.2, gives the careful reader enough information to get started. As soon as you are out in the field with the essence of the process in your head, you will see that there is really little mystery to it.

One member of the team stands on the known position with the pocket transit, while the person with the rod stretches the tape away from the position in a straight line. As the rod carrier moves away, he or she will encounter positions whose elevation it will be necessary to ascertain to complete the map. At these positions the rod carrier stops and sets the rod vertically on the ground. The distance is first taped, and then, with the aid of the level in the transit, the transit reader notes the difference in elevation on the rod. With the transit compass he or she notes the magnetic bearing. All of this should be recorded in a log (notebook) or sketched, as shown in Figure 4.2.

When you can stretch the tape no farther, repeat the process in another direction so that all points are established in a radial pattern from the starting place. When this is finally completed, the transit operator gets a chance to change position by moving to any of the newly established points, and the process begins again. In this manner the contour of the entire property is revealed.

Later, transcribe the readings onto the map, using a protractor and a scale. When it is complete, study the contour map in detail. It will aid in many important design considerations, not only when you begin the homestead but over the years as well. Steepness, drainage, flatness, and exposure all influence land capability, and the map will give a greater understanding of these. Where necessary, the map will guide the efficient modification of the natural contour. Unnecessary earth-moving is extremely expensive.

When the survey map is complete you can begin planning. Much of the homesteading effort depends on the wisdom and efficiency of the plan, the pleasing arrangement of the physical plant—starting with the center of all activity, the hub of the homestead complex, its courtyard. All traffic originates or terminates in the courtyard, so it should have a minimum diameter of 100 feet, in which vehicles may easily turn. Buildings should be readily accessible from the court by the shortest, most direct route possible.

Next, access routes should be designed to connect the courtyard to the other spaces of the homestead. A drive, preferably fewer than 200 feet in length, should link public access with this inner courtyard. The longer this entrance road, the more expensive it will be to build and maintain. Electric and telephone lines are costly when excessive in length. Secondary roads should connect the courtyard with the homestead's component parts—its plant, animal, and human cohabitants. A third access pattern should connect components with each other. A circular route facilitates this, making chore performance more efficient and less tiring, and it eliminates sharp changes of direction and dead ends.

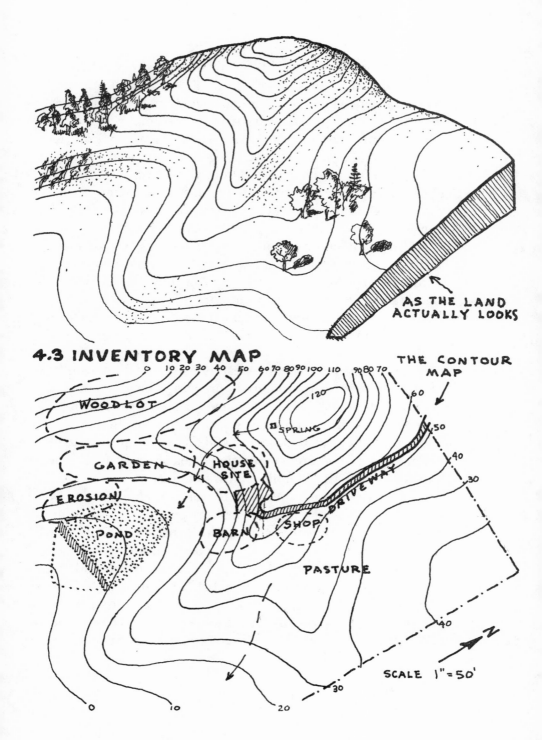

AS THE LAND
ACTUALLY LOOKS

4.3 INVENTORY MAP

THE CONTOUR MAP

WOODLOT

0 10 20 30 40 50 60 70 80 90 100 110 90 80 70

120 60

SPRING 50

40

GARDEN HOUSE SITE 30

DRIVEWAY

EROSION

POND SHOP

BARN

PASTURE

N

40

0

10 30

20

SCALE 1"=50'

Supply buildings should be arranged according to their frequency of use. You should also be able to reach buildings with adjoining yards by truck upon occasion. Construct doorways, gates, and passageways with ample height and width. They should be gently sloped when necessary, provide smooth travel surfaces, and be negotiable by pieces of large equipment where indicated. Planning your work routine will eliminate unnecessary trips. To reduce the number of trips, move maximum loads whenever possible. Use pushcarts or powered equipment for the heaviest loads.

Make flow diagrams evaluating chore routes, work centers, and storage areas, and keep distances between chores at a minimum. Let livestock self-feed. Their buildings and yards should facilitate their ready movement to outlying pasture areas. Since about three-quarters of their feed becomes fertilizer, you must remove manure from animal shelters and holding yards.

In a physical sense a homestead is a complex organism for producing food. Raw materials are assembled, processed, stored, and converted here. This metabolic activity is complicated by factors of climate, space-time relationships, use of equipment, and constant growth and change, requiring an efficient layout. The engineered homestead is a system of organized centers that simplify and systematize work.

Possibly the earliest reference to work simplification as we have come to know it in contemporary production technique came from the Oregon Agricultural Experiment Station in 1931. At that time the station made detailed analyses of factors influencing the cost of egg production. It was found that the most efficient operator expended 1.5 hours per hen per year; the least efficient operator expended 7.3 hours per hen per year. The station's Bulletin 287 concluded: "One of the outstanding causes of lesser efficiency in labor was found to be the chore route . . . in doing the daily chores connected with the enterprise." In subsequent years similar studies were made on dairy farms as well. One made in New Hampshire surveyed thirty-eight farms. The time spent daily on farm chores ranged from 5.2 to 21.1 man-hours.

In 1943 Dr. R. M. Carter made a now-famous work-simplification study under the auspices of the Vermont Experiment Station. For four months Carter studied the work practices of a dairy farmer with a herd of twenty-two milk cows. At this time, with an investment of only $50, Carter rearranged stables, tools, and supplies. He also rearranged milking stalls so that the farmer could travel in a circular route when working with the cows. A new work routine conserved time and travel, reducing milking time from 9 to 4 minutes per cow. New equipment suited to the job at hand was provided. A simple cart used to transport silage reduced the number of trips to the silo from twenty-three to two. Stable-cleaning trips were reduced from seven to three, and a convenient and accessible location was found for tools and supplies. Daily walking distance was, consequently, reduced from 3 miles to 1. Overall chore time was shortened from 5 hours to 3 hours a day. A total of 760 man-hours and 730 miles of walking was saved in one year.

Finally, you must give due consideration to safety and comfort. A healthful, secure environment must be provided for all homestead denizens. Interior shelter should comfort and protect animals in winter, just as it does the homestead family. Animal housing should have adequate daylight and ventilation, especially in warm weather. You can build or plant windbreaks to protect animals and plants from tearing wind and drifting snow. In 1850 Downing noted the relationship between the environment and animal production with these words:

As it is well known now, the extra supply of heat needful for animal economy in cold weather, if it is not supplied by an extra consumption of food with no increase of flesh or strength but with a great loss of comfort to the exposed animal, this extra consumption of food in a few months, even where food is cheap, will more than balance out all that can be saved by withholding a few feet of boards and a few hours' labor.

So far we have considered mostly the physical aspects of homestead-planning. But there are psychological aspects to planning as well. The psychology of space involves a duality: openness and enclosure. The intimacy afforded plants, animals, and humans in their respective shelters—the greenhouse, barn, and house—is essentially provided in what is called the endospace. An arbor, a loafing shed, or a covered patio opens outward from this more or less introverted environment into semiprivate mesospace. And, finally, ectospace is host to all with gardens, orchards, fields, ponds, and woodlot. Aware of the types of space, their functions, and their expressiveness, no one can better articulate the social statement given by a homestead created by its own builders and occupants.

This multifaceted organism is integrated by a utility or nerve system. The shop and garage structures house this maintenance center, which in the larger view includes water and sewer lines as well. Fences, walls, gates, and even culverts are also included in this functioning, as are reservoirs, fuel storage, and a loading platform.

The concept of the homestead as a living, maturing entity requires a homestead plan that is open-ended, allowing for expansion and growth, spatially and actively. Buildings and open yards should be adaptable to change and growth; functions should be flexible and multipurpose.

4.4 HOMESTEAD LAYOUT

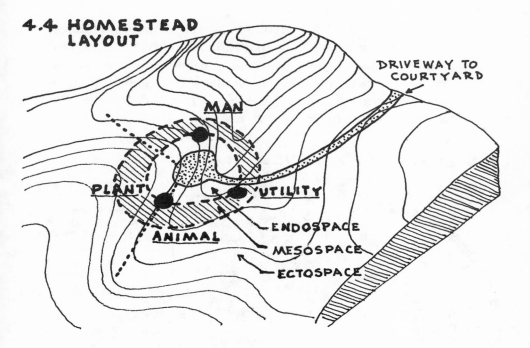

Homestead buildings tend to have a multiplicity of shapes, sizes, roof pitches, window sizes, and floor plans. The best design policy is one that maintains a unifying effect on these various structures.

The layout presented in the photograph in Figure 4.5 is only theoretically possible. Site conditions influence design and arrangement. Solar orientation determines the effects of sun and shade in summer and winter. Wind direction affects the size and location of windbreaks, walls to be insulated, placement of opening windows, and building groupings. Land contour determines water drainage and air flow, which will ventilate not only the family residence but the animal shelter as well. Breeze and view direction should figure into homestead-planning.

More specific information related to planning is offered

further on in this book. Here we intend to inspire and interest you in planning. We hope that sufficient ideas have been included to give the prospective homesteader an overall view of homestead-planning. But inspiration is not enough; fulfillment requires planning. As an old German adage says, "We are too soon old and too late smart."

4.5 HOMESTEAD HOUSE PLAN

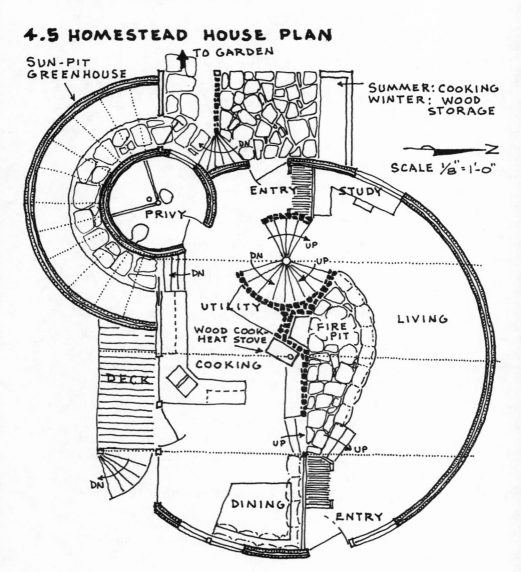

SUN-PIT
GREENHOUSE

↑ TO GARDEN

SUMMER: COOKING
WINTER: WOOD
STORAGE

ENTRY

STUDY

PRIVY

SCALE ⅛"=1'-0"

DN

UP

UP

DN

UTILITY

LIVING

WOOD COOK-
HEAT STOVE

FIRE
PIT

COOKING

DECK

UP

UP

DN

DINING

ENTRY

5

Water Development

*Every human enterprise is the mixture of a little bit of humanity, a little
bit of soil and a little bit of water.*   *Jean Brundes*

Water is the most valuable and indispensable
resource to be found on the homestead. Its proper or improper
development and use will make or break a homestead effort
sooner than any other factor. And yet, for a resource of such
importance, it is surrounded by much misinformation and
even superstition.

Water is our most flagrantly wasted commodity, since, with
air and soil, it is assumed to be one of our most abundant and
resilient resources. It has been estimated that primitive
people's water needs for drinking, cooking, and occasional
washing was 1 gallon a day. By comparison, the average
modern consumer squanders an astonishing 1,200 gallons
each day. (This figure includes a share—approximately 1,000
gallons—of the water used for agricultural and industrial purposes.) Producing 1 ton of steel requires 65,000 gallons of
water; an egg, 225 gallons; and a loaf of bread, 550 gallons of
water. Figure 5.1 shows that fewer than half of the states
experience little or no water deficiency in any season.

How homesteaders set out in search of water is, more often
than not, also an indication of their ignorance of its
availability, use, and conservation. Despite long and well-es-
tablished scientific methods for locating this resource below
ground, many still resort to using what they superstitiously

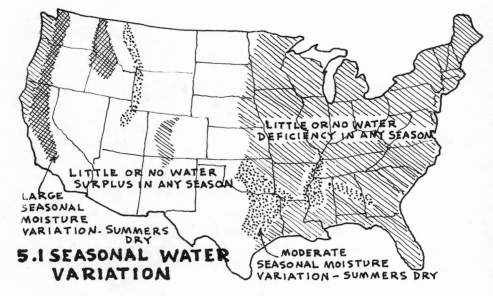

LITTLE OR NO WATER
DEFICIENCY IN ANY SEASON

LITTLE OR NO WATER
SURPLUS IN ANY SEASON

LARGE
SEASONAL
MOISTURE
VARIATION. SUMMERS
DRY

MODERATE
SEASONAL MOISTURE
VARIATION - SUMMERS DRY

5.1 SEASONAL WATER VARIATION

call a divining or dousing rod. Water witches believe this contrivance to be a conduit between electrical currents in their bodies and those of underground streams. Despite the sincerity of some, their method is highly suspect. The New South Wales (Australia) Conservation and Irrigation Commission kept records of all wells drilled from 1918 to 1943. Of 3,581 wells, half were divined and half were not. Of the divined wells, 80.5 percent produced water, while 89.1 percent of the nondivined wells produced water. Some 14.7 percent of the divined wells were "absolute failures," whereas only 7.5 percent, or about half as many, of the nondivined wells failed to produce water.

Examination of the site and ground-water research of your area that can be done at the local library will guide you to properly locating your well. The United States is divided into ten ground-water regions. Each region is further classified by the Water Resources Division of the U.S. Geological Survey, in the Department of the Interior. This agency has amassed

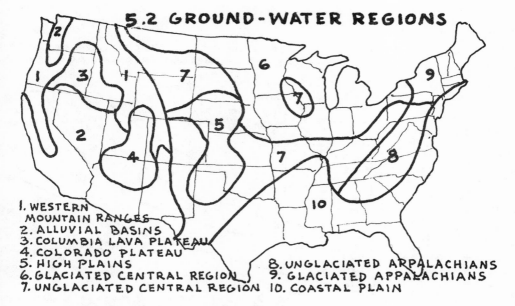

5.2 GROUND-WATER REGIONS

1. WESTERN MOUNTAIN RANGES
2. ALLUVIAL BASINS
3. COLUMBIA LAVA PLATEAU
4. COLORADO PLATEAU
5. HIGH PLAINS
6. GLACIATED CENTRAL REGION
7. UNGLACIATED CENTRAL REGION
8. UNGLACIATED APPALACHIANS
9. GLACIATED APPALACHIANS
10. COASTAL PLAIN

an impressive amount of information on ground-water supply, based largely on geologic study. Much public information also comes from well drillers in the field. Many states have laws regulating the tapping of ground water, and they require drillers to log every well drilled. All of the rock layers penetrated by the drillers and the location, depth, and quantity of water produced must also be shown.

Water-seeking homesteaders should acquaint themselves with the geology of their region. Rocks are the most valuable indicators of *aquifer,* a layer of permeable rock, sand, or gravel that acts as an underground reservoir. Geologic study is helpful for predicting the distribution, depth, and thickness of aquifers. The best is gravel, followed by sand, sandstone, and limestone. In order for a well to be productive, it must penetrate materials saturated with usable water.

Much of this geologic exploration will begin with maps of your site and the general vicinity. Topographic maps from the U.S. Geological Survey indicate lakes and marshes, rivers and

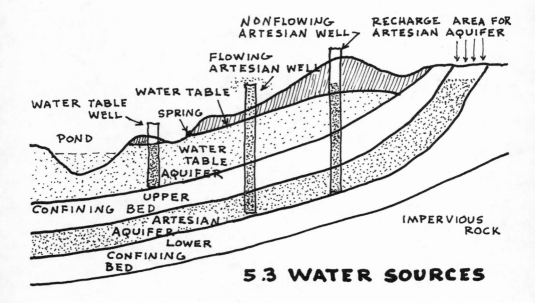

NONFLOWING ARTESIAN WELL

RECHARGE AREA FOR ARTESIAN AQUIFER

FLOWING ARTESIAN WELL

WATER TABLE

WATER TABLE WELL

SPRING

POND

WATER TABLE AQUIFER

UPPER CONFINING BED

ARTESIAN AQUIFER

LOWER CONFINING BED

IMPERVIOUS ROCK

5.3 WATER SOURCES

streams, wells and springs, and geologic folds and faults. A rock body appearing at the surface may serve as a conduit for water below or as a barrier to its movement. Understanding the geologic formation of the region will assist you in identifying and categorizing such features. Weathering, erosion, sedimentation, compaction, volcanism, and glaciation all affect underground water storage capacity.

In general, ground water follows the same movement pattern as surface water. If the sides of an alluvial-filled valley are the same height, water will likely be found in the middle of the depression. If hills flanking one side are higher or steeper, water is more apt to be near the steeper side. Artesian springs can be developed where primary and secondary valleys intersect. The head of a valley is the least likely location for an artesian spring. Plants and trees can be reliable signs of the proximity of ground water, as illustrated in

Figure 5.4. The average depth of all domestic wells in the United States is less than 50 feet. About 90 percent of all ground water lies within 200 feet of the surface.

To estimate your water needs, you must understand climate and the hydrologic cycle. Altitude affects water need; the higher the altitude, the less water growing things need. Cloudy days and rain occurring at heights up to 3,000 feet reduce the evapotranspiration rate on mountain slopes. At these higher altitudes much precipitation also falls as snow and is stored on the ground during winter for use in spring and summer months. A slope facing south will, incidentally, lose its snow as much as 30 days sooner than a slope facing north. Topography has a major effect on water distribution. Slopes facing the prevailing moisture-laden winds, for example, receive more precipitation than leeward slopes.

Without question, a homesteader's first choice is a gravity-fed spring or artesian well. If the spring is at least 20 feet higher than the homestead, moderate pressure will deliver adequate water. Unfortunately, most springs are low producers, requiring some kind of storage facility. However,

5.4 PLANTS THAT INDICATE GROUND WATER

SPECIES	HEIGHT	LOCATION	INDICATES WATER-
RUSHES, COTTONTAILS	GRASS	SWAMPS	NEAR SURFACE
REEDS, CANE	TO 10 FT.	STREAMS	WITHIN 10 FT.
GIANT WILD RYE	6-8 FT.	SUBHUMID	NEAR SURFACE
SALT GRASS	6-12 IN.	SALT FLATS	NEAR SURFACE
PICKLE WOOD	2 FT.	SALT FLATS	NEAR SURFACE
ARROW WEED	SHRUB	THICKETS	10-20 FT.
PALM TREES	50 FT.	ARID	NEAR SURFACE
WILLOW TREES	10-25 FT.	U.S.	NEAR SURFACE
RABBIT BRUSH	SHRUB	WESTERN U.S.	WITHIN 15 FT.
GREASE WOOD	3-6 FT.	U.S.	10-40 FT.
MESQUITE	10-15 FT.	U.S.	10-50 FT.
COTTONWOOD	50 FT.	ARID REGION	WITHIN 20 FT.
ELDERBERRY	30 FT.	U.S.	WITHIN 10 FT.

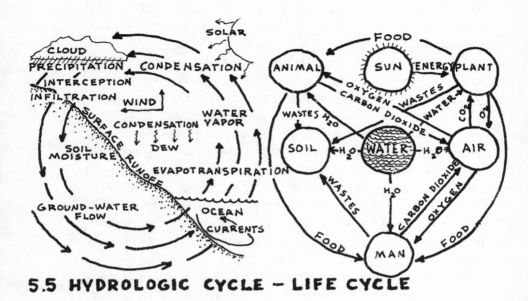

5.5 HYDROLOGIC CYCLE – LIFE CYCLE

with storage a spring emitting as little as 1 gallon per minute will, in a day's time, produce 1,440 gallons of water. This is sufficient to supply forty milk cows and more than sufficient for the average household.

Traditional facilities for water storage are expensive and not durable. A metal tank will rust and a wooden one will rot. Neither is suited for underground installation, where temperature rise and evaporation can be minimized. Unquestionably, the best material to use for constructing a water tank is concrete. A subsurface concrete reservoir will be durable and consistently cool, and in it evaporative loss is minimized.

Ken's earliest contribution to owner-builder homestead development was a low-cost all-concrete circular water tank. Its combination foundation and floor is a single concrete slab.

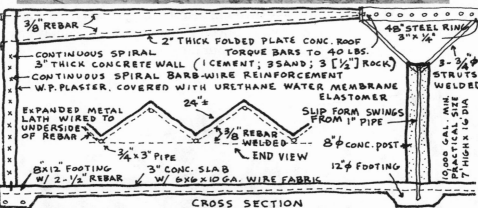

3/8" REBAR

2" THICK FOLDED PLATE CONC. ROOF
TORQUE BARS TO 40 LBS.

48" STEEL RING
3" x 1/4"

CONTINUOUS SPIRAL
3" THICK CONCRETE WALL (1 CEMENT; 3 SAND; 3 [1/2"] ROCK)
CONTINUOUS SPIRAL BARB-WIRE REINFORCEMENT
W.P. PLASTER, COVERED WITH URETHANE WATER MEMBRANE
ELASTOMER

3 - 3/4" φ
STRUTS
WELDED

EXPANDED METAL
LATH WIRED TO
UNDERSIDE
OF REBAR

24" ±

3/8" REBAR
WELDED
END VIEW

SLIP FORM SWINGS
FROM 1" PIPE

8" φ CONC. POST

3/4" x 3" PIPE

8" x 12" FOOTING
W/ 2 - 1/2" REBAR

3" CONC. SLAB
W/ 6x6x10 GA. WIRE FABRIC

12" φ FOOTING

10,000 GAL. MIN.
PRACTICAL SIZE
7' HIGH x 16' DIA

CROSS SECTION

5.6 CONCRETE WATER TANK BUILT WITH A SPIRAL THIN-WALL SLIP FORM

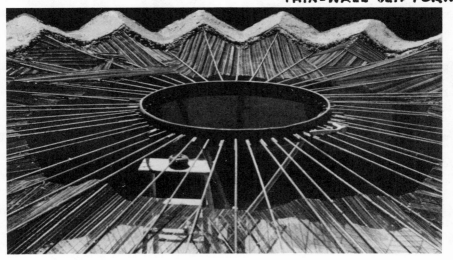

Its roof is a folded plate, a 2-inch concrete pour on expanded metal lath. Three-inch walls are cast using a movable metal slip form. Easily made in the homestead workshop, the slip form is adjustable to any radius and can be used to build a variety of homestead structures. In later chapters we will indicate its use for erecting the walls of garden beds, fish tanks, a compost privy, a greenhouse, a workshop, a silo, and the homestead residence itself.

If you cannot develop a natural spring or an artesian well, you have the choice of digging, boring, driving, jetting, or drilling for water. Each method, discussed briefly below, has its own advantage and application. Your state geological survey office will assist you in determining what type of earth formation you may likely encounter in your area. Simply give them a legal description of your property.

■ UNCONSOLIDATED AQUIFERS
▨ CONSOLIDATED AQUIFERS

5.7 GROUND-WATER AREAS (MAJOR AQUIFERS)

Where the water table is fairly close to the surface a dug well may be practical and economical. Depths of 10 to 40 feet with diameters of about 40 inches are common for these wells. Although square wells have been built, the circular shape is critical for strength and endurance. Except where solid rock is encountered, dug wells require some form of permanent lining. A lining prevents the loosening and slumping of wall material, which would admit contaminated surface water into the well. A lining can also be built to provide support for a pump platform below ground. One unique and practical method for digging deep wells (down to 200 feet) has been developed by the World Health Organization. The first 45 feet of such a dug well is a cast-in-place lining of concrete. Then precast concrete cylinders are assembled and lowered into the well. These act as caissons. As earth is removed with continuing excavation, caissons drop lower, guided by the cast lining above.

Bored wells can also be constructed by hand labor, using a simple earth auger. The maximum practical depth is 50 feet. Usually the auger has a 6- to 8-inch diameter. To bore with an auger, you force the auger blades into soil while turning the tool. When the space between the blades is full of earth, remove the auger from its hole and empty it. As the hole deepens, add lengths of pipe to the auger handle. A pulley-equipped tripod, however, becomes necessary when you reach greater depths. You can thus insert the extended auger rod and remove it from the hole without unscrewing all the sections of pipe. An American Friends Service Committee team in India devised a simple hand-operated boring auger. In place of the tripod they built a 10-foot-high platform. One man perched on the platform could then easily handle pipe lengths of 20 feet. An earth auger functions best, incidentally, in heavy clay soil.

In coarse sand it may be better to drive a well. Driven wells are usually 2 inches in diameter and less than 30 feet deep. If driving conditions are good, you can drive a 4-inch casing as deep as 50 feet. A driving tool consists of a drive point connected to the lower end of sections of pipe. A drive point is a perforated pipe with a steel head, which breaks through earth as it strikes. Five-foot lengths of pipe are used for the pipe "string" and serve as the well casing when driving is completed. Driving is done by either of two methods. Strike a drive cap, which covers and protects the upper end of the casing, with a maul or by a weight falling from a tripod and guided by the casing. Or, again using the tripod, allow a steel driving bar, attached by a rope to a pulley on the tripod, to fall freely inside the casing, striking the drive cap.

A jetted well is created by drilling into the earth with a high-velocity stream of water produced by a jetting pump, which delivers 50 to 100 gallons per minute at 50 pounds of pressure. A self-jetting well point washes fine particles upward out of the casing, which sinks by its own weight as ground is washed from beneath it. Install a tripod hoist to support the drill pipe and casing. It can also be used to drop a drive weight on the pipe when penetrating clay soil. Rotate the casing so that the teeth at the lower end will cut into the earth at the bottom of the hole. Use a straight bit to penetrate hard formations that will not yield to the water jet. You can mechanize the operation by using a portable gasoline engine to rotate an earth- or rock-cutting bit.

You must drill wells where you encounter stone or hard formations, and it is best to engage commercial well-drilling rigs for this work. Use either the cable tool percussion method or the hydraulic rotary method. With the former method, shape the well with the pounding and cutting action of a chisel-type drill bit by alternately raising and dropping it.

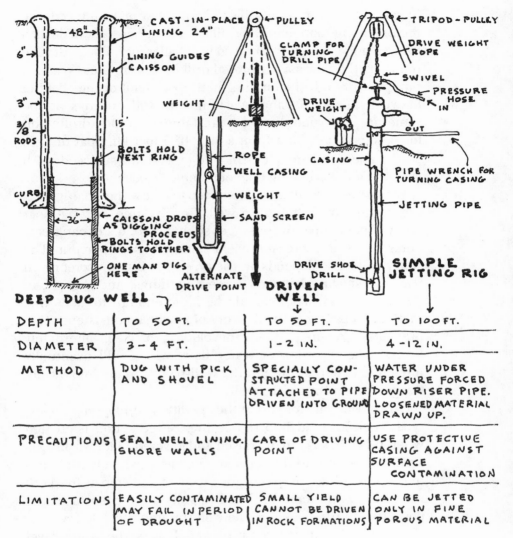

	DEEP DUG WELL	DRIVEN WELL	SIMPLE JETTING RIG
DEPTH	TO 50 FT.	TO 50 FT.	TO 100 FT.
DIAMETER	3-4 FT.	1-2 IN.	4-12 IN.
METHOD	DUG WITH PICK AND SHOVEL	SPECIALLY CONSTRUCTED POINT ATTACHED TO PIPE DRIVEN INTO GROUND	WATER UNDER PRESSURE FORCED DOWN RISER PIPE. LOOSENED MATERIAL DRAWN UP.
PRECAUTIONS	SEAL WELL LINING. SHORE WALLS	CARE OF DRIVING POINT	USE PROTECTIVE CASING AGAINST SURFACE CONTAMINATION
LIMITATIONS	EASILY CONTAMINATED MAY FAIL IN PERIOD OF DROUGHT	SMALL YIELD CANNOT BE DRIVEN IN ROCK FORMATIONS	CAN BE JETTED ONLY IN FINE POROUS MATERIAL

5.8 HAND MADE WELLS

Suspend the bit from a cable. When water is introduced, the reciprocating motion of the drilling tool mixes loosened earth into a sludge, which is removed by a bailer tool. The hydraulic rotary method uses a perforated drill bit that rotates. Water is pumped through the bit and then up and out of the opening between bit and casing, washing out drill cuttings at the same time. In your shop you can fabricate a rock-drilling bit for percussion drilling from a 3-inch, 5-foot-long bar of mild steel. The 90-degree cutting edge is hard-surfaced with stellite. Devise a bailing bucket to remove loose soil.

Well development is the next step to follow the driving, boring, jetting, or drilling of a well. Remove fine material near the well point, opening passages so that more water can freely enter the well, increasing yield as much as 50 percent. The water yield also depends on the type of well-point screen you use, the spacing between screen openings, and the size of those openings. They should be large enough to allow fine material to pass into the delivery pipe with a consistency that prevents clogging. When developing a well, alternately reverse the direction of water flow, which will eventually force fine material through the well-point screen, into the delivery pipe, and out of the well.

One device for this part of the operation employs a plunger-type block lowered into the casing. Surge action is created when this block is rapidly raised and lowered. Backwashing is another method used to develop further an already functioning well. With a pump designed to start and stop at frequent intervals, water can be lifted rapidly to the surface and permitted to run back into the well through the delivery pipe. This intermittent action raises and lowers the water level through screen openings.

Well tests follow well development. Test the water yield to ensure that the proper pumping equipment will be installed.

Pump until you have determined the number of gallons delivered per minute. Ten gallons per minute is considered adequate for homestead use. This test not only establishes the well capacity in gallons per minute but also determines the recovery rate of the water level after pumping has ceased. You can also determine water "hardness" at this time. Excessive amounts of calcium and magnesium decrease the lathering capacity of common soap. Figure 5.9 shows the areas in this country where hard ground water might be found.

Before discussing the various pumping systems that require some external power source, we should mention the hydraulic ram, a pumping system powered by the water it pumps. Although in use since 1800, the ram is still little known, in both principle and application. Yet you can use a

UNDER 120
120-180
180-240
OVER 240

HARDNESS AS CaCO₃
PARTS PER MILLION

5.9 HARDNESS OF GROUND WATER

ram where you either have or can develop a source of flowing water amounting to at least 3 gallons per minute with a fall of no less than 3 feet. Under ideal conditions this nearly perfect pump will lift water as much as 500 feet from its source. You can figure on getting about 25 feet of lift for every foot of fall. For its operation the water falls through a drive pipe, gaining speed until it forces an automatic valve to close abruptly. This suddenly arrested flow creates pressure, forcing some of the water past a check valve and into an air chamber. The air in the chamber is thereby compressed and finally forces the water up the delivery pipe via an exit at the bottom of the air chamber. Usually the terminus is a storage tank from which you can draw water for homestead use.

Once in operation, a ram will require only minimal maintenance for thirty or more years of constant service. Hydraulic rams are available from manufacturers in this country but you can also fabricate one in the homestead workshop for a fraction of its commercial price. The Intermediate Technology Group (9 King Street, London, England) has devised a simple ram applicable for use in producing a homestead water supply. Nothing but pipe fittings and materials commonly found about the homestead shop is used.

Pumping systems for wells are generally divided into two groups—those for shallow and those for deep wells. Of the former, the oldest and most common is the piston pump. It is simple in design, will deliver 250 to 500 gallons per hour, and is especially suited to wells of low capacity. A piston or a plunger working back and forth inside a cylinder creates suction.

Where winds average at least 5 miles per hour for more than half the required pumping time, a windmill can be used to power a piston pump. The mill should tower at least 15 feet above surrounding buildings and trees. The size of the mill is determined by the depth of the well. Blades 6 feet in diameter (from the tip of one blade to the tip of its opposite member) will pump from depths of 70 feet; 8-foot blades, from 140 feet; and 10-foot blades, from 250 feet.

The most efficient shallow well pump is the rotary gear, which consists of two gears that mesh inside a housing. The revolving gears capture water between gear teeth and housing and, as they continue to revolve, squeeze the water up and out through the delivery pipe. The supply of water is continuous and steady, without pulsation, and water pressure is high.

Centrifugal, or throwing-action, pumps will deliver large quantities of water with great efficiency. However, they lack the positive action characteristic of the piston and rotary-gear pumps; that is, as the water level drops, a centrifugal pump draws proportionately less water.

Deep-well pumping is somewhat more complex and involves putting the suction part of the pump, if not the entire pump, either into the water or close to its surface. A simple type of deep-well pump is the piston pump, which operates on the same principle as it does in shallow wells. The major difference is that the cylinder is attached to the delivery pipe and lowered into the well to within 25 feet of the water level. The piston or plunger, still inside the cylinder, gets its motion via a sucker rod connected to the power source at ground level. Again this arrangement is suited to wells of continuous, but low, production—with delivery the same as in a shallow well.

Probably the most common deep-well pump found on the modern homestead is the centrifugal jet pump, which is a centrifugal pump with a nozzle added. As water passes through the nozzle its speed is greatly increased. A diffuser or venturi-shaped opening converts the high speed into high pressure by the pump. The jet, consisting of nozzle and

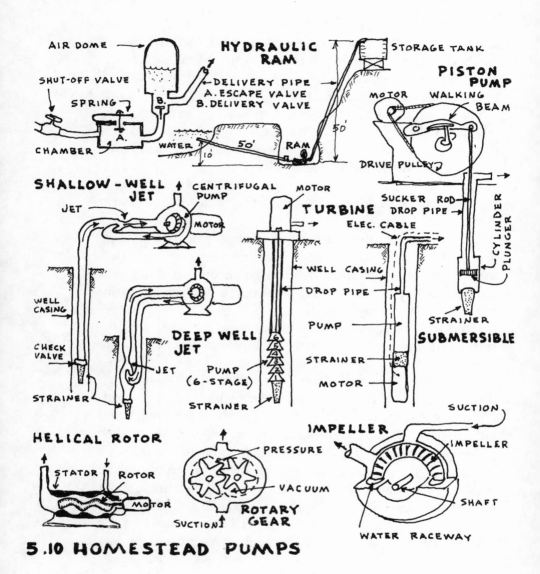

AIR DOME

SHUT-OFF VALVE

SPRING

CHAMBER

HYDRAULIC RAM

DELIVERY PIPE
A. ESCAPE VALVE
B. DELIVERY VALVE

WATER   50'   RAM

STORAGE TANK

PISTON PUMP

MOTOR   WALKING BEAM

DRIVE PULLEY

SUCKER ROD
DROP PIPE

CYLINDER PLUNGER

SHALLOW-WELL JET

JET

CENTRIFUGAL PUMP

MOTOR

WELL CASING

CHECK VALVE

STRAINER

DEEP WELL JET

JET

MOTOR

TURBINE

ELEC. CABLE

WELL CASING

DROP PIPE

PUMP

STRAINER

MOTOR

PUMP (6-STAGE)

STRAINER

SUBMERSIBLE

STRAINER

SUCTION

IMPELLER

IMPELLER

SHAFT

WATER RACEWAY

HELICAL ROTOR

STATOR   ROTOR

MOTOR

PRESSURE

VACUUM

SUCTION

ROTARY GEAR

5.10 HOMESTEAD PUMPS

venturi, is located below water level, which should not exceed 100 feet. This is a high-capacity pump under low pressure.

Below 100 feet a submersible pump is commonly used. This is a centrifugal pump and motor assembly, built as a watertight unit completely submerged in the well. The direct coupling of motor and pump gives greater efficiency and its immersion effectively cools the motor. The unit is suspended on the end of the delivery pipe and has its own strainer. The only parts above ground are the control box and the pressure switch.

Another deep-well pump worthy of discussion is the turbine, a vertically mounted multistage centrifugal pump. Any number of impellers can be added to raise water from any depth. The series of impellers is mounted on the low end of the delivery pipe in the water. Through the delivery pipe runs a shaft that supplies power from the motor at ground level.

Figure 5.11 shows the relative merits of various pumps. This table is intended to aid homesteaders in choosing the proper pump. Water needs will be discussed in the next chapter.

Naturally a homesteader's first choice for water development would be the simpler, maintenance-free water system, one that may be developed and installed with minimum equipment and plumbing. Here we must discuss the new and innovative horizontal well. Developed in the early 1950s with equipment originally engineered for highway drainage, this method has the potential of bringing gravity-fed water systems to many homesteads lacking artesian wells, springs, or deep wells.

Two types of sites are suited to horizontal-well development: (1) a geologically tilted formation, which will sometimes create a rock dike (or dam) for underground water and which can be tapped by drilling through the impervious bar-

	RAM	PISTON	ROTARY	IMPELLER	JET	TURBINE
COST	LOW	LOW	MODERATE	MODERATE	MODERATE	HIGH
CAPACITY	10 GAL./HR.	10-25 G.P.M	10 G.P.M.+	2 G.P.M. TO UNLIMITED	6-130 G.P.M.	25 - 5000 G.P.M.
HEAD	HIGH	HIGH	HIGH	2 - 150'	LOW	7-150'
OPERATION	SIMPLE	SIMPLE	SIMPLE	SIMPLE	SIMPLE	DIFFICULT
EFFICIENCY	LOW 5%	LOW 25%	GOOD 50%	GOOD 50%	LOW 40%	HIGH 85%
LIFT	UNLIMITED	TO 600 FT.	22 FT.	10 FT.	TO 85 FT.	TO 1000 FT.
ADVANTAGES	SIMPLE DESIGN, LITTLE ATTENTION, WATER FOR POWER	POSITIVE ACTION, WIDE SPEED RANGE, USE HAND OR WIND POWER	POSITIVE ACTION, STEADY DISCHARGE, WIDE SPEED RANGE	QUIET, CAN USE HORIZONTAL, GOOD FOR LARGE VOLUMES	MOVING PARTS ON SURFACE, QUIET, EASY PRESSURE HOOKUP, NEED NOT BE SET OVER WELL	QUIET, NO VIBRATION
DISADVANTAGES	WASTES WATER, NOISY, MUST BE PERMANENT OPERATION	MUST BE SET OVER WELL, VIBRATION NOISY	SUBJECT TO ABRASION, NOISY	NOT GOOD FOR LOW CAPACITY, MUST BE SET NEAR WATER	JET NOZZLE SUBJECT TO CLOGGING	REQUIRES LARGE-BORE WELL, REPAIRS REQUIRE PULLING
MAINTENANCE	NONE	PLUNGER REQUIRES ATTENTION	LITTLE	LITTLE	SIMPLE. ATTENTION REQUIRED	DIFFICULT. SKILL REQUIRED

5.11 PUMPS AND POWER

SERVICE LIFE

ELECTRIC	DIESEL	GAS (AIR COOLED)	GAS (WATER COOLED)	PROPANE
50,000 HRS. 25 YRS.	28,000 HRS. 14 YRS.	8,000 HRS. 4 YRS.	18,000 HRS 9 YRS.	28,000 HRS 14 YRS.

rier to the aquifer, and (2) a perched water table, which is sometimes found above an impervious layer of rock. At the point where water seeps out, near the top of the impervious layer, a horizontal well can be drilled to the aquifer.

Horizontal drilling is a sensible way to develop a natural seep or spring. Too often, digging or blasting to expose aquifer will destroy the natural barrier—the overburden—

which serves to dam the underground reservoir. In a sense a horizontal well is a cased spring, containing water within a closed system from its point of origin to its point of use. For this reason it provides a sanitary water supply, while with an open spring contamination is common and difficult to control.

The process of horizontal-well development involves a small rotary-jet rig with a pipe chuck powered by a 5-horsepower gas engine and a water pump capable of delivering 3 gallons per minute at 120 pounds of pressure per square inch. Standard 1¼-inch steel pipe is used as a drill stem. The bit has tungsten carbide blanks welded into notches in the leading edge. As the drill moves at about 100 revolutions per minute, the chuck and stem move forward on a carriage, while water pumped through the stem cools the bit and removes cuttings. To avoid vacuum problems in the casing, a minimum downward slope of ½ inch per foot is recommended. Heavy clay, decomposed granite, and soft rock can

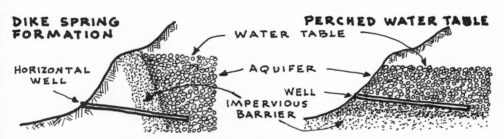

DIKE SPRING FORMATION

HORIZONTAL WELL

WATER TABLE

AQUIFER

IMPERVIOUS BARRIER

PERCHED WATER TABLE

WELL

5.12 HORIZONTAL-WELL DRILLING

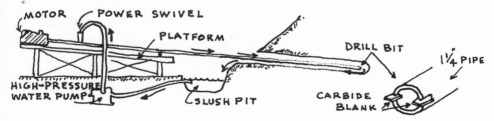

MOTOR   POWER SWIVEL

PLATFORM

HIGH-PRESSURE WATER PUMP

SLUSH PIT

DRILL BIT

CARBIDE BLANK

1¼ PIPE

be drilled at a rate of 3 to 9 inches per minute and hard rock at 1 inch per minute. A diamond bit is used in extremely hard rock. Horizontal wells up to 200 feet in length have been successfully drilled.

A low-cost, simplified horizontal-drill rig can be devised with an earth attachment for a power chain saw. Stihl of West Germany is one company that manufactures a highly efficient and reliable chain saw with such an earth-drill attachment. This power arrangement creates little waste or duplication, since the chain saw is an indispensable tool on the homestead.

6

Water Management

Afric's barren sand,
Where nought can grow, because it raineth not,
And where no rain can fall to bless the land,
Because nought grows there.   *Marsh*

Like land and things that grow on it, water has been squandered. To develop arid regions in which streams flow only during unproductive seasons, we go great distances to divert water from courses originating in wetter climes. Impounding this captive water behind large dams, we overirrigate once-infertile soil, causing nitrate salts to be leached from the ground and pathogenic fungi to proliferate there. We then apply commercial fertilizers and poisonous sprays to soil and crops to counteract the evils inflicted by our profligate use of water.

On the other hand, when farming rich bottomland we drain adjacent hills, valleys, and meadows of their stored water. To perpetrate this rape of a resource, we install tile drainage systems, lowering the water table and causing downstream flooding. We cut or burn forests and native grasslands in order to graze cattle, or we plow the land to replace native vegetation with crops whose cultivation depletes the soil. Tillage practices leave the land exposed to the ravages of wind and rain. Agriculture thus becomes, for the most part, an occupation fostering erosion and infertility, floods and drought, insect infestations, and diseased plants.

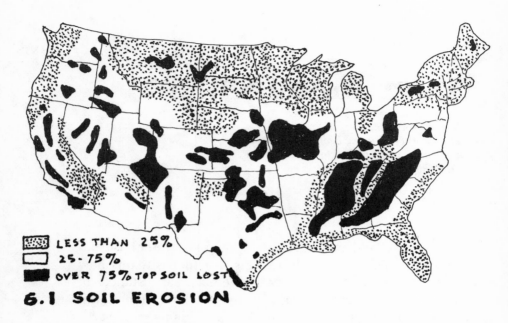

LESS THAN 25%
25-75%
OVER 75% TOP SOIL LOST

6.1 SOIL EROSION

Egypt's Aswan High Dam is one striking example of modern man's disregard for the principles of the hydrologic cycle. The dam has disrupted the water supply to the Nile by reversing the course of aquifers feeding the river. At its estuary, fish have been deprived of the flow of nutrient-laden silt and are rapidly disappearing from long-established fishing grounds in that area of the Mediterranean. And as a result of the river's dormancy, eroding silt barriers in the Nile Delta threaten freshwater lagoons with the invasion of seawater.

Vast areas of desert the world over have recently been supplied with irrigation water. As a result, the water table has risen, carrying crop-destroying salts and minerals to the surface. Occasioned by this increased salination, land that had been uncultivated for thousands of years because of drought again lies abandoned. The Aswan High Dam fiasco illustrates that water and water management are inseparably linked to land and the way it is used.

When the homesteader understands water use in its relation to growing plants and living soil, an efficient water-management program can be initiated. Rain (or irrigation water) forces used air out of soil and pulls fresh air in as it seeps downward. Rain water, having absorbed carbon dioxide in its passage through the atmosphere, is slightly acid. The acidity dissolves minerals and other nutrients, rendering them assimilable by plant roots. The nutrients then combine with those the plant gets from the air through its leaf structure.

A healthy soil, as defined in the following chapter, absorbs water quickly, preventing runoff with its accompanying erosion. Having reached its optimum saturation point, generally considered to be two parts water to three parts soil by volume, a healthy soil will absorb no more. This relationship of water to soil is conducive to vigorous plant growth and should give you some idea of the kind of loose, open structure a soil should have. When this optimal two-thirds saturation point is reached, excess water is stored in the subsoil, where it will not evaporate and will be available when needed.

The first principle, then, of proper water management is to increase subsoil water storage. Water in subsoil depths reaches plant leaves—where food production takes place—by capillary action through plant roots. The importance of subsoil water storage is obvious when the sheer volume of water required by plants is considered. For instance, wheat, beans, and peas use about 200 pounds of water for every pound of dry matter produced. The healthier the soil, the less water required. It has been found that manure applied to infertile soil will actually reduce the water required to produce 1 pound of corn from 500 gallons to 300 gallons. Fertilizing is also essential for the extensive and vigorous root development necessary for the effective utilization of subsoil moisture.

Plants also use water to regulate temperature, in that a moist leaf absorbs heat to vaporize moisture. To avoid overheating, a plant will expend energy to transpire moisture. Midday sprinkling on a hot, sunny day will increase photosynthesis. The cooling effect reduces plant energy expended to transpire moisture, thereby increasing energy expended in cell growth. The timeworn advice that you should sprinkle only during cool evening hours is just one more disproven adage.

Organisms and plant roots in the soil inspire carbon dioxide and respire oxygen. This ingress and egress are requisite for healthy plant growth. Irrigation-flooded land weakens crops by impeding this transpiration process. Flooded soil also promotes undesirable bacterial growth. Harmful anaerobic (water-borne) microorganisms thrive, reducing valuable nitrates to toxic nitrite and gaseous nitrogen, which readily escapes into the atmosphere.

Tillage operations further impede plant growth. Heavy equipment compacts soil and creates another obstacle to nutrient absorption by obstructing root growth. As tillage equipment stirs soil, all sides of the soil particle are exposed to air, permitting moisture to evaporate. Even though part of the purpose of tilling is to destroy weeds, which would otherwise compete for moisture, no more moisture is available to a cultivated crop than to a crop sown on untilled, mulch-covered land; and this is true even of light tilling. We deplore the effort that is wasted on dust mulching. Once a dry layer is formed on the surface of soil, no amount of hoeing or cultivating can appreciably reduce the rate of moisture evaporation. Cultivation also destroys surface feeder roots. For other reasons, too (detailed elsewhere in this text), we view the hand-operated rotary tiller as the most undesirable piece of equipment employed on would-be organic homesteads.

Mulch planting can significantly limit surface evaporation. We will say more about this method in ensuing chapters, but to the extent that mulch saves water and protects soil, it should be touched upon here. Mulch planting is the cultivation of new crops in the residue of a previous crop without prior land preparation. Crop residues left on the surface of the soil throughout the winter help conserve water by their absorption of moisture and their accumulation of snow, which might otherwise be blown away.

Mulch-planted crops require less water and fewer applications, because water infiltration is greater and there is less runoff. Organic matter acts as a soil cover and reduces evaporative losses. Table 6.1 illustrates the runoff and the resulting erosion experienced in conventional farming.

Table 6.1: Soil Loss under Various Treatments (Ohio, 1969)

Average Annual Soil Loss

Rotation-plowed cornland, sloping rows	7 tons/acre
Rotation-plowed cornland, contour rows	2 tons/acre
No-tillage mulched cornland, contour rows	Trace
Wheatland	1 ton/acre
Meadowland	Trace

Soil Loss under Severe Test (5 inches of rain in 12 hours, July 5, 1969)

Rotation-plowed cornland, straight sloping rows	22 tons/acre
Rotation-plowed cornland, contour rows	3 tons/acre
No-tillage mulch cornland, contour rows	0.03 tons/acre

It does not take long farming experience in a semiarid region to appreciate fully some basic concepts of water management. A homesteader should attempt wise control of as many other environmental growth factors as possible so that water becomes the only limiting factor. You should have seed in (or, when mulching, *on*) the ground by the earliest safe planting date. This will enable seedlings to benefit from accumulated winter moisture while the evaporation rate is relatively low. Crop sequence is also important. Shallow-rooted crops, such as corn, do not tap available subsoil moisture; deep-rooted crops, such as grains, do. It is prudent to follow corn with grain or alfalfa to benefit from moisture untapped by the corn.

The competition among plants for available water is fierce. Small plants, for instance, have little chance for survival when they are planted too close to tree crops. In very dry seasons expand the distance between plant rows to increase crop yields. This conserves water during the early part of the growing season, making it more readily available as the crop approaches maturity.

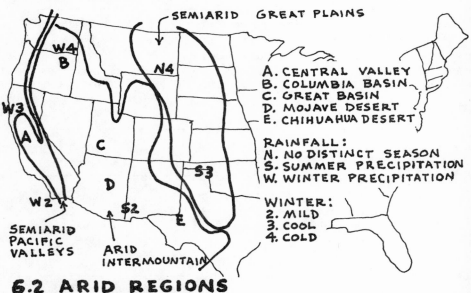

SEMIARID GREAT PLAINS

A. CENTRAL VALLEY
B. COLUMBIA BASIN
C. GREAT BASIN
D. MOJAVE DESERT
E. CHIHUAHUA DESERT

RAINFALL:
N. NO DISTINCT SEASON
S. SUMMER PRECIPITATION
W. WINTER PRECIPITATION

WINTER:
2. MILD
3. COOL
4. COLD

SEMIARID PACIFIC VALLEYS

ARID INTERMOUNTAIN

6.2 ARID REGIONS

6.3 WATER CONSERVATION

Another water-conserving technique, contour farming, is becoming widely practiced on sloping land. Planting on contour reduces the velocity of runoff, conserves water, and impedes erosion by encouraging thorough infiltration of the soil and percolation to subsoil depths.

Fallowing is a common practice where the distribution of rainfall is sparse and uneven. Land left uncultivated for a growing period stores water for the next season's crop; that is, one crop is grown from moisture received in a two-season period. Fallowing is inefficient, but it does facilitate water storage in rain-poor country.

Indiscriminate use of irrigation is a Pandora's box of problems associated with soil management—soil structure, aeration, drainage, leveling, organic content, mineral deficiency, toxicity, salt accumulation, and infertility. As we pointed out earlier, water management has a subtle but far-reaching influence! Human tampering can easily upset the hydrologic balance. Mindless irrigation, especially in arid regions, will destroy the soil profile, making plant propagation futile. The salt in irrigation water will never evaporate. It accumulates in amounts toxic to plants. About one-quarter of all irrigation water is used to wash salt out of the growing zone. But the salt is eventually deposited in the subsoil below the root zone or in irrigation ditches, where it must be reckoned with. Excessive irrigation also leaches out nutrients, leaving soil severely deficient in nitrogen and phosphates.

The extent of the damage depends upon the structure and texture of the soil. Heavy, clayey soils hold more water with less nutrient-leaching. The structural aggregates of heavy soils retain nutrients but allow water to drain around them. Light soils are extremely sensitive to excess water.

Drip irrigation, the newest method of crop-watering, conserves water and soil. This system delivers small quantities of water applied frequently but at a slow rate by a low-pressure water system. As a result, orchard trees reach bearing age in one-third the time of those conventionally watered. "Water stress," the shock that plants experience after being alternately drenched and parched, does not occur with drip irrigation nor are harmful salts deposited in the growing zone. Less drying-down between waterings keeps salts more dilute. Drip irrigation also minimizes weed growth, since only 10 percent of the soil in the root zone of newly planted crops is wetted. For native crops this ratio may be increased to 33 percent, still considerably less than with conventional methods. According to the proceedings of the Second International Drip Irrigation Congress, drip irrigation of a mature orchard took half as much water as was used in conventional irrigation. This saving is primarily due to the fact that evaporation losses are minimized because the water applied at a slow rate is fully absorbed in the root zone.

A drip-irrigation system consists of emitters, lateral lines, main lines, and controls. Emitters control the water flow from laterals to the soil, allowing water to emerge as droplets. One gallon per hour is the usual emission rate, although some models are manually adjustable. Lateral and main lines usually are small-diameter plastic pipes laid in fairly level terraces to facilitate low operating pressure (10 pounds per square inch). Controls filter the water and regulate pressure, volume, and period of application.

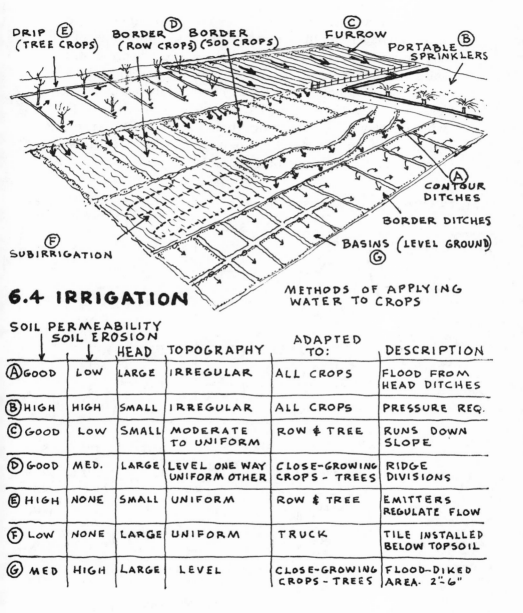

DRIP Ⓔ
(TREE CROPS)

BORDER Ⓓ BORDER
(ROW CROPS) (SOD CROPS)

FURROW Ⓒ

PORTABLE Ⓑ
SPRINKLERS

CONTOUR Ⓐ
DITCHES

BORDER DITCHES

BASINS (LEVEL GROUND)
Ⓖ

Ⓕ
SUBIRRIGATION

6.4 IRRIGATION

METHODS OF APPLYING
WATER TO CROPS

	SOIL PERMEABILITY	SOIL EROSION	HEAD	TOPOGRAPHY	ADAPTED TO:	DESCRIPTION
Ⓐ	GOOD	LOW	LARGE	IRREGULAR	ALL CROPS	FLOOD FROM HEAD DITCHES
Ⓑ	HIGH	HIGH	SMALL	IRREGULAR	ALL CROPS	PRESSURE REQ.
Ⓒ	GOOD	LOW	SMALL	MODERATE TO UNIFORM	ROW & TREE	RUNS DOWN SLOPE
Ⓓ	GOOD	MED.	LARGE	LEVEL ONE WAY UNIFORM OTHER	CLOSE-GROWING CROPS - TREES	RIDGE DIVISIONS
Ⓔ	HIGH	NONE	SMALL	UNIFORM	ROW & TREE	EMITTERS REGULATE FLOW
Ⓕ	LOW	NONE	LARGE	UNIFORM	TRUCK	TILE INSTALLED BELOW TOPSOIL
Ⓖ	MED	HIGH	LARGE	LEVEL	CLOSE-GROWING CROPS - TREES	FLOOD-DIKED AREA. 2"-6"

It is difficult for laypersons to estimate how much water a given plant will require. For one thing, the properties of the soil in which the crop is grown have a direct bearing on the soil's affinity for moisture. The coarser the particles, the less water they will hold. Sandy soils have a moisture capacity of less than 1 inch of available water to a depth of 1 foot, whereas clay soils retain more than 3 inches of water to the same depth. In a vegetable garden an equivalent of 1 inch of rain is considered adequate in a single downpour. Most

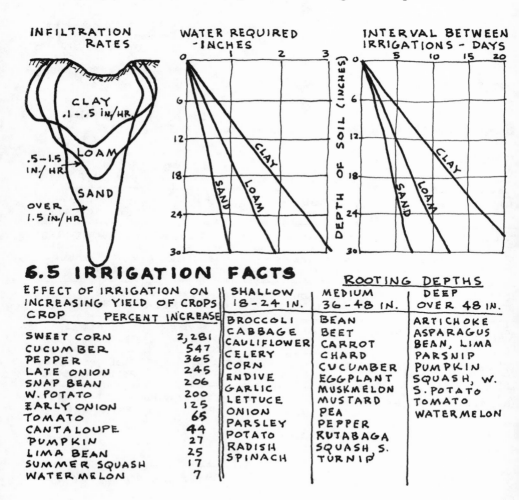

6.5 IRRIGATION FACTS

EFFECT OF IRRIGATION ON INCREASING YIELD OF CROPS		SHALLOW 18-24 IN.	ROOTING DEPTHS	
CROP	PERCENT INCREASE		MEDIUM 36-48 IN.	DEEP OVER 48 IN.
SWEET CORN	2,281	BROCCOLI	BEAN	ARTICHOKE
CUCUMBER	547	CABBAGE	BEET	ASPARAGUS
PEPPER	365	CAULIFLOWER	CARROT	BEAN, LIMA
LATE ONION	245	CELERY	CHARD	PARSNIP
SNAP BEAN	206	CORN	CUCUMBER	PUMPKIN
W. POTATO	200	ENDIVE	EGGPLANT	SQUASH, W.
EARLY ONION	125	GARLIC	MUSKMELON	S. POTATO
TOMATO	65	LETTUCE	MUSTARD	TOMATO
CANTALOUPE	44	ONION	PEA	WATERMELON
PUMPKIN	27	PARSLEY	PEPPER	
LIMA BEAN	25	POTATO	RUTABAGA	
SUMMER SQUASH	17	RADISH	SQUASH S.	
WATERMELON	7	SPINACH	TURNIP	

vegetables require 0.1 to 0.2 inches of water a day, or about 3 to 9 inches a month. A garden crop of 0.1 acre requires 3,000 gallons of water over the entire season. Research at the Kansas Experiment Station disclosed some interesting facts about crop-watering. Wheat seeded in dry soil yielded 5 bushels per acre. Soil wet to a depth of 1 foot yielded 9 bushels; to 2 feet, 15 bushels; and to 3 feet, 27 bushels.

In cases where land is plentiful but water is scarce, it has been found more efficient to moisten a large area with a small amount of water than to flood-irrigate a small area with a large amount of water. A 4-acre crop receiving an application of 7 inches of water yields more than three times as much as a 1-acre crop receiving 30 inches.

A small farm pond is an essential part of an owner-built homestead system. As a reservoir it is ideal for any method of crop-watering. It can also be used for watering livestock and raising fish, ducks, and geese. And the fringe benefits are many—beauty, summer and winter recreation, increased land value. It can also serve as a plentiful water source in case of fire.

The history of pond-building is rife with unsuccessful attempts, the reasons for failure being many and varied. While it is not within the scope of this book to go into the engineering details of small-dam construction, some of the obvious criteria for choosing a pond site should be mentioned.

Seek a site that offers the maximum amount of storage area for the smallest dam practicable. Ideally, the site should be located at a point higher than the homestead environs to deliver gravity-fed water to the home, service buildings, and croplands. A steep-sided valley with a slowly running stream is ideal. The greater the cross-sectional rise, the narrower the dam at its top. The flatter the longitudinal section, the farther upstream water will be impounded.

Closely investigate the geological strata of your dam site. A concrete dam is the practical choice where outcroppings of granite or basalt extend across the valley near the surface of the ground. Locate the dam wherever good rock formation is

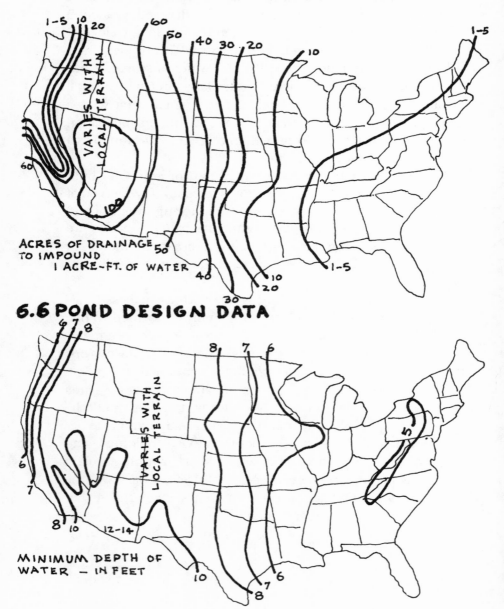

ACRES OF DRAINAGE
TO IMPOUND
1 ACRE-FT. OF WATER

6.6 POND DESIGN DATA

MINIMUM DEPTH OF
WATER — IN FEET

found. Avoid loose rock outcroppings along stream banks. Sometimes a shallow layer of soil will cover objectionable gravel or shattered rock. For an earth dam the site should contain a large percentage of clay and some silt or sand. Too much clay causes cracks when dry and slippage when wet. Too much sand causes seepage.

An adequately designed spillway is critical to pond management. The purpose of the spillway is to carry surplus water from the pond, away from the face of the dam. It may consist of a mechanical control, such as an exit pipe installed in the base of the dam where it will empty below the dam site. Or it may be an open drainage ditch planted to some protective ground cover, situated clear of the dam face (to one side of it), and cut into solid ground whence water flows onward down the valley.

Discourage trees, shrubs, and deep-rooted legumes from growing on the dam, since water tends to course along their root channels and may thereby endanger the stability of the dam. Trees planted around the pond should be coniferous—not broad-leaved varieties—to minimize water evaporation through their palmate leaves. Positioning the trees correctly will reduce evaporation losses by sheltering the pond from winds that might otherwise ruffle the surface of the water. Wave action manifoldly increases water evaporation. Waves and rain also erode pond banks.

Erosion from rain may be a fitting subject with which to end this chapter. Erosion is like pollution; it indicates grave mismanagement of the environment on a scale so vast that it is almost incomprehensible. At fault is the whole thrust of present-day agricultural practice. Rains, for instance, are heaviest during that part of the growing season when crops are least able to provide ground cover. This fact alone should suggest that you adopt a different practice for planting, one that allows the simultaneous growth of weeds and crops.

Farsighted agriculturists have found that when weeds are allowed to grow with corn, the crop yield increases. Weeds are not the water robbers that people once thought them to be. Their value as conditioners and protectors of the soil far outweighs the cost of their support. Weed growth loosens soil and enlarges the feeding zone of other crops. Crop roots will follow weed roots deep into the subsoil in search of water.

Proper water management requires that soil be used within its capacity. A homestead ought to be sectioned into cropland, pastureland, woodlot, and wildlife refuge, depending upon the structure of the soil, its depth, its slope, the movement of air and water through it, and its susceptibility to erosion by wind and water.

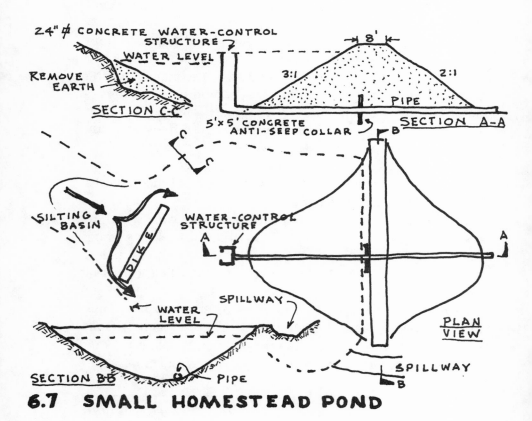

6.7 SMALL HOMESTEAD POND

7

Soil Management

I am done with great things and big things, great institutions and big success. I am for those tiny invisible molecular forces that work from individual to individual, creeping through the crannies of the world like so many soft rootlets, or like the capillary oozing of water, yet if given time, will rend the hardest monuments of men's pride.   *William James*

As in the larger world we inhabit, where barely perceptible forces sometimes wreak changes of magnitude, so in the world of microbes tiny forces, if given time, will render plant nourishment from inert, inorganic resources in soil— the repository for animal and human sustenance as well. Everything in agriculture starts with the lowest common denominator: the final crop is a product of its relationship with the soil colloid, at which level those "tiny invisible forces" work, exchanging nutrient-producing materials. The whole range of activity—from colloid to crop—comprises a system that involves soil structure, tilth, biotic activity, and nutrient uptake. These compose the physical, biological, and chemical aspects of the story of soil.

As a medium for plant growth, soil is a complex natural material derived from disintegrated rock and decomposed organic materials, all of which provide nutrients, moisture, and air to plants. The development of soil from its original parent rock materials is a long-term process involving both physical and chemical weathering along with biological activity. Parent rock is cracked and chipped by temperature change. As it is broken, there is an increase in the surface area of rock

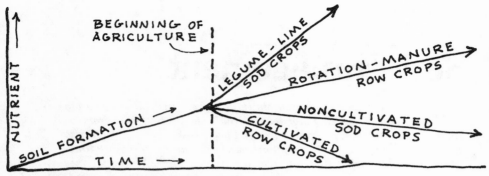

7·1 SOIL EVOLUTION

particles exposed to the atmosphere. The chemical action of water, oxygen, carbon dioxide, and various acids further reduces the size of rock fragments and changes the chemical composition of the resulting materials. Finally, the action of microorganisms and higher plant life contributes organic matter to the weathered rock material and a true soil begins to form.

To understand and appreciate the character of soil, examine a number of soil samples from the homestead cropland. If samples are healthy, soil granules, or crumbs, will be aggregated into a structural unity. Disaggregated soil has no structure. Instead, its subsurface pores are small and unstable. Such soil dissolves into mud with the first rain and afterward dries to a powdery dust. Only when you understand the intricacies of soil structure can you appreciate earth's ability to produce edible, nourishing plants.

Ideally structured soil has large and stable pores extending from surface to subsoil. The size and arrangement of soil particles govern the flow and storage of water, the movement of air within the aggregate, and the environment for nutrient exchange between soil and plants. A ready supply of air is

SOIL STRUCTURE DESTROYED BY CULTIVATION

FROM VIRGIN LAND—EXCELLENT STRUCTURE

RESULT OF SOAKING, AGGREGATES REMAIN INTACT IN SOILS HAVING WELL-DEVELOPED GRAIN STRUCTURE

COMPARING LIGHT- AND DARK-COLORED SOILS

7.2 SOIL SAMPLES

especially necessary so that organic matter may be decomposed by aerobic bacteria. Spaces around soil particles also act as channels for conducting water to root hairs. About half the volume of soil should consist of soil pores and fissures. These spaces contain microorganisms that form gums and mucilages as they break down, helping to bind soil particles together. Therefore, maximum pore space is essential for robust plant chemistry and biologic activity.

Pore spaces in soil aggregates must be contiguous from subsoil to atmosphere. A photomicrograph would show root formation grown in sandy soil as it is illustrated in Figure 7.3. Note how the clay coating around grains of sand virtually becomes an extension of the root cell itself. This is the world of the soil colloid, which is made up primarily of inorganic materials, such as clay, sand, highly decomposed organic materials, and humus. It is here that the most vital and intimate environment for nutrient exchange between soil and living plants is found.

Soil microorganisms are composed of microscopic living animals, called microfauna, and microscopic plants, known as microflora. They are in contact with almost every particle

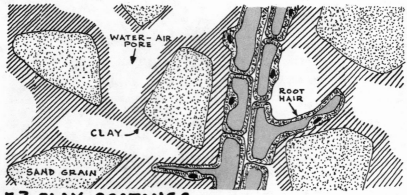

7.3 CLAY COATINGS

of soil, whereas plant roots are not. They churn earth into top-soil rich in minerals. Their digestion neutralizes both alkalies and acids in the soil. Earth minerals and chemicals are broken down, enriching the soil with nutrient particles that seedlings and plants can assimilate. Without microorganic life, soil would become an inert mass incapable of providing food.

Of the microbes in soil, 99 percent are bacteria. The balance are fungi, protozoa, and algae. Soil bacteria are of two kinds: aerobic, which thrive in conditions of abundant oxygen, and anaerobic, which live in the absence of oxygen or where pore spaces are filled with carbonic acid and water, as in heavy or hardpan soils. The activities of these microbes are prodigious. One variety of aerobe uses air from pore spaces to produce a compound of nitrogen for plant protein-building. Other equally interesting microbes live for mutual benefit on the feeder roots of certain plants. This symbiotic relationship aids plants in the assimilation of minerals needed for their development.

The success of soil management depends on the care and feeding of microbes. Plants produce organic matter, which is returned to the soil where these decomposers use it as their source of nutrients. In so doing the microbes provide the inorganic and simple organic nutrients required by plants for the synthesis of the more complex organic compounds. In this mutually beneficial relationship, therefore, organic materials must be recycled back into the environment from which they came. In nature, relationships are cyclical and all cycles are interrelated.

In any discussion of soil populations earthworms are always prominently included. Actually, in comparison with soil microorganisms, earthworms, being much higher in the food chain, bestow benefits more limited in their effect. Darwin overstressed the case for the ubiquitous worm. At best, earthworms are indicators of soil fertility, not its cause. The earthworm has no mechanism for creating exceptional plant food, for capturing solar energy, or for fixing nitrogen from the air. Earthworm movement creates insignificant soil aeration, and there is no unusual amount of richness to be gleaned from its castings. The leafy diet of the earthworm is especially low in mineral content, and to the extent that the earthworm feeds on the organic content of humus, it burns energy and reduces soil fertility.

As vegetable matter is returned to the soil it is digested by microbes. The resulting cellular material is mixed with the bodies of bacteria, fungi, and other microscopic forms of life, together with certain excretory materials produced during their life cycles, to form a dynamic, ever-changing, organic matter called humus. (We could even call the nutritive layer of the earth a humusphere.) Humus is at once both the source of energy and food for soil organisms and a storehouse of plant essentials. The breakdown products and dead bodies of microbes give humus its characteristic dark color.

Microbes in the humus layer slowly decompose organic matter, liberating continuous streams of soluble carbon dioxide, which combine with water to form carbonic acid. Parent subsoils thereby release essential minerals, and nitrogen and phosphorus are obtained from air in pore spaces. This medium fosters brisk air circulation and a capacity for moisture retention. As a result, humus absorbs more heat and releases it more readily. However, to prevent the literal burning up of humus, more organic matter must continually be supplied, especially to sandy soil.

Humus is an amazing material that has qualities and functions as yet unknown to modern science. One pound of the colloids in humus is said to cover 5 acres of surface area. S. A. Waksman, in his major work on humus, mentions that when all extractions are made by known processes, 30 percent of the humus is unaccounted for. Most important to an understanding of how humus is formed is the concept that growth equals decay. The growing season is also a season for decay.

Composting is basically a controlled process to produce humus. Mulching, however, is an equally viable process of soil enrichment, as mulch requires no further handling once it is spread directly on the ground. Organic matter, applied as mulch, slowly decomposes without the excessive heat buildup and the loss of nutrient gases that take place in the composting method. A program of composting with mulch is called, appropriately, sheet composting.

In our judgment planting in mulch is the best alternative to tilling soil. Practical techniques for planting various crops in this manner are discussed in following chapters. Our interest at this point is in how mulch aids in soil structure and fertility. Obviously, covering the land with decaying organic fiber prevents compaction and crusting of the soil surface. Beneath this surface the decay process continues unabated. Research at the New Jersey Agricultural Experiment Station has disclosed that fresh organic residues applied directly to soil on an equal basis produce as much as three times the soil aggregate as did compost prepared from identical materials. The resulting colloidal medium acts as a conduit for air and water, connecting the upper layers of soil with subsoil nutrients and preparing the way for the penetration of plant roots. A list of the benefits to be derived from the application of organic matter follows:

1. Soil will warm faster, enabling you to work the ground and plant earlier; plants will emerge and mature more quickly, allowing you to harvest crops sooner.
2. Soil will become friable, taking and holding water better and reducing irrigation needs by about one-third.

3. Weeds will become easier to control.
4. Disease and insect infestation, nature's indicators of the unfit, will become less of a problem.
5. Preparation for planting will be minimized.
6. Soil will be built up, and plants will be provided with nutrients as they need them.
7. Crops will be vigorous, healthy, and of high yield.
8. You will benefit by the knowledge that diminishing resources and energy reserves are not unnecessarily being used up, as is the case when petrochemical fertilizers are used. Nor will you be adding to pollution but rather cooperating with ecological cycles.

The foundation of productive land is its organic content. Force roots, not tillage tools, through soil.

To ensure that we will have an adequate supply of suitable, high-quality food from productive land, we must add only materials for soil enrichment that we can go on using forever. For present and future generations of mankind we must employ a program of food production as nearly self-generating as possible. Commercial fertilizers, made from nonrenewable fossil fuels and mineral deposits, fail to do this. And by synthesizing inorganic and simple organic plant nutrients, we prematurely retire soil microbes from their indispensable activities, destroying their populations and producing dead soil.

These decomposers have been around supplying nutrients to plants for eons. We are novices in our attempts to fecundate the earth—as seen by the frequency with which our efforts contaminate water supplies, increase crop disease and pest attack, and poison vegetables with toxic levels of nitrate that cannot be eliminated naturally. By merely dumping quantities of manufactured nutrients into soil, we fail to feed plants what they need. Whatever form of enrichment you add to the soil, the principle of soil replenishment remains the same: given sufficient organic content, microbe populations will select for themselves, individually, only those nutrients compatible with their specific needs. We would ensure our survival by cooperating with these decomposers and by offering them a balanced diet.

Conventional agriculture, as it is practiced today, is concerned only with simple solutions for easy profit. Its concern with single communities of plants (monoculture) is hopelessly shortsighted. The most stable communities are the most complex, being made up of a large number of species. Therefore, any program of soil enrichment for such a varied group must be carefully considered. William Albrecht, one of the world's most thoughtful soil scientists, tells us that the goal of fertilization ought to be the arrangement of exchangeable mineral ions into their respective percentages. Obviously, this is not going to happen with profit-minded farming interests, which produce only single, cash crops. Besides the usual provision for the acid soil elements of nitrogen, phosphorus, and potassium, a homesteader must provide for the careful apportionment of the base soil elements of calcium, magnesium, sodium, and certain trace elements.

Since nature is inclined toward balance, there is an electrical attraction between acid elements of soil colloids and base (mineral) elements of the subsoil, resulting in the vital exchange of these elements on the colloid level. Besides this attraction and exchange, there is chemical-electrical bonding, which assures the delivery of components for the eventual synthesis of food for plants. Higher living forms exist by the dynamics of soil chemistry.

When this sequence is short-circuited, chemical fertilizer

may perform rescue chemistry, and it may help for a while, although its use must ultimately be paid for. The heavy, indiscriminate, or continual use of commercial fertilizer can have a detrimental effect on plant growth. Phosphate rock, for instance, is treated with sulfuric acid to make it more readily available to plants. As soon as plant roots reach these soluble chemicals they go into spasmodic growth, which in turn upsets the delicate balance of hormones and enzymes in the growing plants, making the plants magnets for insect and fungal attack. The solubility of commercial fertilizers is much too high in most instances, and it is therefore too rapidly assimilated. These fertilizers also supply soluble salts to newly planted seeds, which have, by nature, their own stored organic provisions. This overstimulates the growth of stems and leaves and interferes with seed development in mature plants.

Unlimited quantities of minerals are available only from soil colloids that have the proper aggregate formation. Mineral nutrients are locked into these aggregations, becoming unavailable for plant nutrition in soils that are compacted, waterlogged, or deficient in moisture. Most soils tend to become acidic as a result of the acid produced by the decomposition of plant materials and the acidic character of some types of fertilizers. The acidity of soil is often expressed in terms of its hydrogen-ion concentration as its pH, which stands for "parts of hydrogen." This is measured on a scale from 4, highly acid, to 10, highly alkaline. Raising the pH of an acid soil from 5 to 6, making it more alkaline, increases the availability of phosphorus about ten times. Reducing the pH of overly alkaline soil from 8 to 7 increases by 500 percent the availability of phosphorus. Most plants thrive at a pH range of from 6 to 7. In this range bacteria thrive, speeding the decay of organic matter and the liberation of nitrogen in the useful

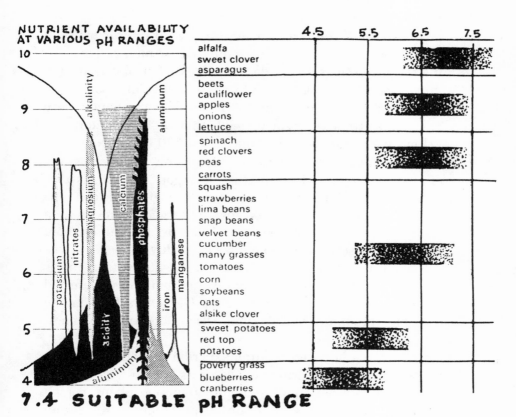

NUTRIENT AVAILABILITY
AT VARIOUS pH RANGES

7.4 SUITABLE pH RANGE

form of nitrates, from which plant proteins are built. Below a pH of 6, only the bacteria that break down organic matter into an inferior ammonia are active.

A parent soil containing large amounts of quartz, granite, sandstone, or shale usually produces acid topsoil, whereas marble and limestone parent soils produce alkaline topsoil. Ground limestone (calcium carbonate) is customarily applied to soil in an effort to reduce acidity. If it is available, use dolomite limestone, since it contains magnesium carbonate as well as calcium carbonate. Magnesium is an important soil amendment. If soil is too alkaline, apply sulfur or gypsum. It is the oxidation of sulfur that reduces alkalinity, and organic matter encourages the oxidation process. This illustrates the importance of organic matter to soil. A light, sandy soil or one that is highly weathered requires less soil amendment to raise or lower pH than a clay soil or one high in organic content.

Figure 7.5, taken from Bulletin No. 140 of the New Zealand Department of Agriculture, shows the varying quantity of limestone required for the three categories of soil, assuming all have an initial pH of 5. Clay soil of a given pH contains more exchangeable ions of positively charged hydrogen than sandy soil of the same pH. Consequently, it requires more limestone (base element) to reduce its acidity.

Since the more acidic portion of a soil is its surface, this is the best place to apply limestone. Yet farmers have preferred to disc it into the lower root zones. Gardeners tend to use either too much or too little lime. Discretion and a soil audit are called for.

Most states have agricultural experiment stations and extension services that will test soil for its lime and its fertility. However, such a simple organic-matter, phosphate, potash soil test, indicating pH and liming requirements, provides no

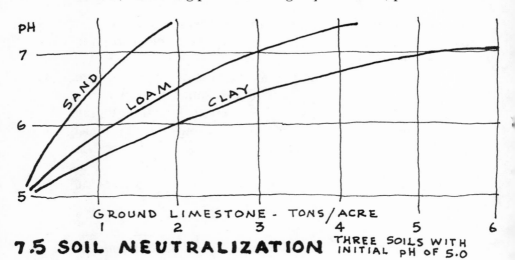

7.5 SOIL NEUTRALIZATION THREE SOILS WITH INITIAL pH OF 5.0

scientific basis for the evaluation of the equally important base soil elements. It is just as important to know the percentage of calcium, magnesium, and potassium in soil. A complete soil and plant-tissue test kit is available from the Agricultural Experiment Station at Purdue University in Lafayette, Indiana, at a nominal charge for those who prefer to do their own soil-testing.

NITROGEN	PHOSPHORUS	CALCIUM	MAGNESIUM	POTASSIUM
SLOW GROWTH DECREASE IN GREEN COLOR YELLOWING SMALL STEMS LESS BRANCH- ING	SMALL PLANTS LOW GRAIN YIELD DELAYED MATURITY SHRUNKEN PODS AND GRAIN	WHITE LEAF SPOTTING STUNT GROWTH DROUGHT OR WINTER-KILL WILTING	STREAKED LEAVES YELLOW-GREEN MOTTLING LEAF SHED PURPLE-RED LEAVES	SCORCHING OF LEAF EDGE SHORT JOINTS IN GRASSES CUPPED LEAVES IN LEGUMES

7.6 SOIL DEFICIENCY SYMPTOMS

Take a core sample for testing from the upper 6 inches of soil where most crop roots feed. It is important to keep soil moist several days before testing, as drought affects pH by killing bacteria. Cold soil also inactivates bacteria. To get an accurate reading of pH, test warm, moist soil. When doing your own testing, remember that components vary from season to season. In early spring, for example, nitrogen is monopolized by soil organisms, giving a false reading. A single soil sample will not necessarily be typical of the entire homestead. A farm usually has three to six different types of soil.

There are tens of thousands of soil types in the United States alone, but their classification is misleading and of little practical value. Early in the history of farming in this country, before the destruction of topsoil and humus, areas (other than desert) that had loam were composed of identical layers of thick black organic matter. It is only since we have lost this

7.7 SOIL-CLIMATE-VEGETATION

TEMPERATURE INCREASE → (vertical)			

ARID DESERT SHRUBS AND GRASSES			
TROPICAL DRY	WARM DRY	COOL DRY	
LIGHT - COLORED SOILS			

SEMI-ARID SHORT GRASSLANDS			
TROPICAL DRY	WARM DRY	COOL DRY	
BROWN SOILS			

SUB-HUMID TALL GRASSLANDS			
TROPICAL	WARM	COOL	
BLACK PRAIRIE AND CHERNOZEM SOILS			

HUMID FORESTS			
TROPICAL WET	WARM HUMID	COOL HUMID	
RED SOILS	YELLOW	GRAY-BROWN	PODZOLS

WET RAIN FORESTS			
TROPICAL WET	WARM HUMID	COOL HUMID	
RED SOILS	YELLOW	GRAY-BROWN	PODZOLS

PRECIPITATION INCREASE →

great resource that the necessity to classify the underlayers of soil has arisen. But, since our approach is to rebuild that resource rather than continue to rape what remains of it, the scientific name for the soil we start with is insignificant.

The most important thing homesteaders should know about the soil they work is whether its texture is clayey, silty, sandy, or gravelly. As noted in a previous chapter, water percolates rapidly through light, sandy aggregations. Soils of clayey texture have greater water-storage capacity. The finer texture a soil possesses, the more nutrients it will have because of the clinging surfaces of its colloidal structure. Water and nutrients are easily leached away from sandy-textured soil. The sandier the soil, the less organic matter it contains and the more necessary it is to build up its humus content.

Inasmuch as soil is created primarily by the growth of plants, the best method to improve soil is to grow crops on it.

Soil aggregations of heavy clay or silt particles can thereby be lightened or made friable by growing weeds or "green manure." Green manure is an improvement crop grown to be eventually returned to the soil. Besides providing surface mulch, roots from green-manure crops penetrate deep. Their

7.8 ENVIRONMENTAL INTERACTION

SOILS

ANIMALS ⇄ PLANTS

CLIMATE

SOIL	PROFILE	NATIVE VEGETATION	CLIMATE	NATURAL FERTILITY	DOMINANT AGRICULTURE
(ARID) CHERNOZEM	2-4 FT. BLACK SOIL TO WHITISH CALCAREOUS	TALL-GRASS PRAIRIE	TEMP. TO SUB HUMID	VERY HIGH	WHEAT- CORN
CHESTNUT	1-3 FT. DARK BROWN TO WHITISH CALCAREOUS	MIXED TALL & SHORT GRASS	TEMP. ARID	HIGH	WHEAT - GRAZING
PEDOCAL — BROWN	1-2 FT. BROWN TO WHITISH CALCAREOUS	SHORT- GRASS PRAIRIE	TEMP. ARID	HIGH	WHEAT- EXTENSIVE DRY-LAND FARMS LIMITED CROPS
SIEROZEM	1 FT. GRAYISH GRADING INTO LIGHT-COLORED CALCAREOUS	SHORT-GRASS & DESERT	TEMP. ARID	MEDIUM	GRAZING
DESERT	GRAYISH SOIL	DESERT PLANTS	TEMP. ARID	MEDIUM	GRAZING
PEDALFER — PODZOL	GRAY LEACHED SOIL OVER BROWN	CONIFER FOREST	COOL MOIST	LOW	SMALL FARMS CROPS & LIVESTOCK
GRAY-BROWN	GRAY-BROWN LEACHED OVER BROWN HORIZON	DECIDUOUS FOREST	COOL MOIST	MEDIUM	SMALL FARMS CROPS & LIVESTOCK
(HUMID) PRAIRIE	3-5 FT. DARK BROWN TO LIGHT PARENT	TALL-GRASS PRAIRIE	TEMP. MOIST	HIGH	SMALL FARMS MUCH LIVESTOCK
YELLOW	GRAY-YELLOW SOIL OVER YELLOW	FOREST (CONIFER)	WARM MOIST	LOW	SMALL FARMS TRUCK & FRUIT
RED	YELLOW-BROWN OVER YELLOW	FOREST (DECIDUOUS)	WARM MOIST	MEDIUM	SMALL FARMS WIDE CHOICE

decomposition leaves a water conduit through which excess water can drain and alongside of which the roots of the next crop of edible plants can easily grow.

Very poor, eroded, structureless soil can be made productive by practicing green-manuring. First, apply the necessary lime and then use a commercial nitrate fertilizer. The reclamation of very poor soil and the making of explosives are the only legitimate uses for manufactured fertilizer. Next, turn the land over to weed growth. Weeds will thrive on the applied nutrients and will penetrate the subsoil with their highly developed root systems. Their roots, in turn, will bring mineral nutrients to the upper soil. Some grasses and legumes will, however, thrive on poor soil and are excellent for green-manuring. These are buckwheat, rye, lespedeza, and sweet clover.

There are other farming practices, such as crop rotation, that improve soil structure. A discussion of these practices brings us to the subject matter of the following chapters on crop management.

8

Sod Crops

You ask me to plough the ground. Shall I take a knife and tear my mother's bosom? Then when I die she will not take me to her bosom to rest. You ask me to dig for stones! Shall I dig under her skin for bones? Then when I die I cannot enter her body to be born again. You ask me to cut grass and make hay and sell it and be rich like white men. But how dare I cut my mother's hair?   *Chief Smohalla*

      The Nevada chief Smohalla's reply to agricultural directives from Washington is understandable. From a proud tradition of nomadic living in harmony with their surroundings, the Nevada people faced a moment of truth. While we would moderate the Indian way, we do respect the reverence with which the Indians held the body of our mutual great host, earth. We, as offspring, stewards, and guests, should manifest a comparable, thoughtful reverence.

Grasses have covered the earth for a long time. Improved, cultivated varieties of alfalfa were known to the Persians and Romans before the time of Christ. Clovers were grown in northern Europe in A.D. 800. There are at present 5,000 species of sod grasses in the world; 1,500 of these are found in the United States. This number does not include the hundreds of native and weed varieties that agriculturists reject as not valuable for animal pasturage.

Without question, sod crops are the most important underdeveloped agricultural resource in the world today. Yet because of their numbers, they are difficult to classify. Some are grown annually, some in rotation, and some permanently.

Of the annuals, there are warm-season and cool-season varieties. Grasses are often grown companionably with other grasses or legumes. Each has its optimum requirement for soil and climate. A high-yield sod crop, such as alfalfa, may prove valueless when grown on poor soil. Conversely, the palatable short grasses (for example, buffalo blue gama, silver bluestem, and sand dropseed) are considered undesirable for production in good soil but may yield excellent results on less productive land.

The case for sod crops has been stated by an eminently qualified agriculturist, A. T. Semple, who began his investigations into the improvement of pasture for livestock in 1916, working for the Department of Agriculture. In recent years Semple has been at work for FAO of the United Nations and the OAS on both sod crops and livestock improvement in seventeen countries. A lifetime's work is summarized in his book *Grassland Improvement*. Although Semple maintains his personal bias for sod crops, he advocates integrating them with animals and other crops and has called for a system of food production particularly applicable to the homestead economy.

Today, we realize more about the interdependence of plants and animals and about the effect it has on the food chain, soil enrichment, plant and animal well-being, and the potential for superabundant crop yields. It is known, for example, from research conducted in Australia and New Zealand that the mere presence of animals in a pasture improves the quality of protein produced on that soil, and the benefit extends beyond that supplied by the usual deposits of animal excrement. Agriculturists have no official explanation for this curious result (although it may well be influenced by carbon dioxide levels), but they state unequivocally that an important interaction takes place between the crop and the grazing animal.

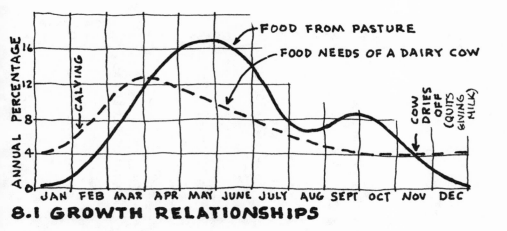

ANNUAL PERCENTAGE

FOOD FROM PASTURE

FOOD NEEDS OF A DAIRY COW

CALVING

COW DRIES OFF (QUITS GIVING MILK)

JAN FEB MAR APR MAY JUNE JULY AUG SEPT OCT NOV DEC

8.1 GROWTH RELATIONSHIPS

It was found, furthermore, that with long-term rotation, where the land was grazed continuously for eight to ten years before intensive cropping, plant and animal relationships improved. The customary short-term rotation of one or two years of grazing followed by a crop of wheat and a fallow period does not achieve the same beneficial results. There is a precedent in history. Twelfth-century Scottish monks realized the benefits of long-term crop rotation and successfully instituted an eight-year plan, which has been in use on monastery land for over six hundred years. The monks also realized that planting legumes stimulated the growth of grass plants by adding a growth factor (nitrogen) to the soil.

The successive development of various species of sod crops is apparent when overgrazed or overcultivated land is abandoned and grass is allowed to return. First, annual weeds (such as crabgrass, pigweed, and Russian thistle) appear. Then the poverty grasses, wire grass and broom sedge, replace annuals. As the organic content of the soil begins to build, poverty grasses get so thick that they cannot withstand their own competition. This provides an opportunity for short-

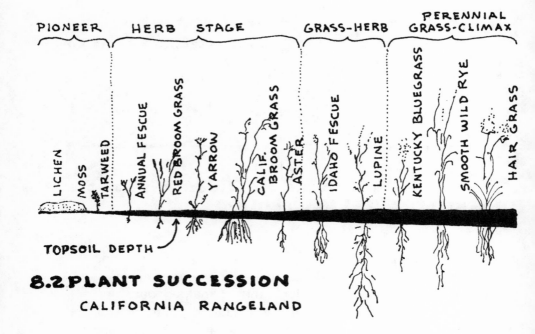

PIONEER   HERB   STAGE   GRASS-HERB   PERENNIAL GRASS-CLIMAX

LICHEN · MOSS · TARWEED · ANNUAL FESCUE · RED BROOM GRASS · YARROW · CALIF. BROOM GRASS · ASTER · IDAHO FESCUE · LUPINE · KENTUCKY BLUEGRASS · SMOOTH WILD RYE · HAIR GRASS

TOPSOIL DEPTH

8.2 PLANT SUCCESSION
CALIFORNIA RANGELAND

lived perennials, such as bunchgrass. These perennials progress in their development from modest beginnings to what is
called climax growth, producing grasses said to be native to
the region. An example of this growth on California rangeland is illustrated in Figure 8.2.

Climax growth is the result of a process in which several
species of plant appear in a given location. Because of the
plants' response to subtle environmental changes, some
varieties of a species are gradually replaced by other, more
highly adapted, varieties. Eventually, certain of the species
disappear altogether, while those that remain emerge as the
ultimate form of vegetation for that particular soil, location,
and climate. Each successive crop contributes to the next
evolutionary development of the species.

Sod attains its ability to support climax growth as more and more topsoil is formed and accumulated. When native sod is plowed and put into production, climax vegetation is destroyed, although the soil may remain productive for some years. When a forest is removed, however, soil becomes unproductive within a few short years after cultivation.

The importance that weeds play in the successive development of optimum sod cover cannot be overstated. Weeds condition soil. They germinate well in the anaerobic conditions of hardpan. Their vigorous root systems open soil, pulverizing it and enlarging the feeding zone for the benefit of subsequent plants. Grass will return to a pasture only after weeds have first prepared the way. When the land is fallowed, grass has the propensity to disperse weeds once environmental conditions are conducive to its return.

The director of the Rocky Mountain Forest and Range Experiment Station, Dr. David Costello, studied the value of weeds on western range. He found, surprisingly, that weeds made up the greater portion of the diet of cattle grazing there. Perennial weeds and shrubs were found to have a higher content of crude protein than native grasses.

It is better for animals to graze grass pastures that are low in protein in summer, when food values are highest. They should graze pasture with an abundant complement of weeds and shrubs in late summer, fall, and winter, when the food value of grass is lower. The nutrient value of standing (unharvested) grass drops during the fall and winter. At that time nutrients drain back into the root zone, while the food value of browse plants (weeds and shrubs) remains high.

You can easily revive abandoned pastureland by first applying manure and lime and then planting a mixture of grass and legume seed, such as rye and alfalfa. Rye is the hardiest of all grasses and is injured least by insects and disease. After applying rock phosphate, cut the grass for two years and leave it to rot back into the soil.

Nitrogen is the most important of the fertilizing elements required for plant growth. It directly regulates a plant's ability to make protein. Therefore, the early stages of plant growth require large amounts. Much usable nitrogen is lost to the atmosphere unless its salts are held in the root zone in solution. For this reason, farmers of former centuries left fields in rough stubble throughout winter months, making it easier for snow and rain to penetrate the ground. In this way they minimized the loss of nitrogen to the atmosphere.

All nitrogen comes from the air. It is returned there, at the same constant rate that it is removed, by plant and animal metabolic processes. The same process makes phosphorus and other elements available to plants. Field sorrel, for example, is a heavy phosphorus user, yet a chemical analysis reveals no phosphorus present in soil in which it grows.

Often a homestead grass pasture requires only renovating, introducing a legume into an existing crop of grass. You usually should renovate a pasture in the early spring, after closely mowing or grazing the crop the previous fall. The close cropping weakens the grass stand, reducing its capacity to be competitive with the newly introduced legume. Test the soil before seeding and apply any lime dressing that might be required. Do not use nitrogen fertilizer, for it will increase grass competition with the renovative crop. Seed with a vigorous legume, such as red clover. When grass begins to shade the legume seedling, allow livestock to graze the field until they begin to graze the legume. Concentrate livestock in order to graze down quickly. Then "turn off," or remove livestock from, the field and allow the pasture to recover. Pastures may require several years to develop adequate root systems.

In a later chapter we show how the food value of row crops depends primarily upon the large quantity of protein and vitamins they contain. Except for the legume family, sod crops cannot accumulate large quantities of nitrogen. Therefore you should follow a sod crop of mixed grass and legume with a ley crop. A ley crop is a pasture sown to annual root crops, such as turnips or potatoes. The practice of alternating row crops with sod crops grazed by animals is called ley farming, a practice that fully utilizes the fertility cycle. The program calls for a three-year sod crop rotated with three years of row crops. Ley farming may provide the soil enrichment desired without the tilling required by green-manuring.

Although it takes about 7 acres of sod crop to equal the production by volume of 1 acre of row crop, you can grow sod crops at a fraction of the cost of row crops. The Bureau of Dairy Industries made a four-year study of the labor cost of producing 100 pounds of completely digestible nutrients. It was found that from pasturage the return per man-hour was six times higher than from corn and ten times higher than from barley.

PERCENTAGE EMERGENCE OF VIABLE SEED AFTER 25 DAYS WHEN PLANTED AT DEPTHS OF:

CROP	0	¼"	½"	1"	0	¼"	½"	1"
	NOT MULCHED				MULCHED 1 T/AC*			
ALFALFA	42	75	63	48	76	85	82	73
RED CLOVER	40	44	39	25	74	85	86	70
ALSIKE "	39	21	22	7	70	91	91	46
LESPEDEZA	48	88	82	33	70	96	92	66
ORCHARD GRASS	37	58	59	40	78	95	86	69
TIMOTHY	30	46	27	3	49	70	68	23

*ONE TON/ACRE

AVERAGE AMOUNT OF NITROGEN FIXED /ACRE

ALFALFA	194#LBS
SWEET CLOVER	119
KUDZU	107
LESPEDEZA	85
VETCH	80
PEAS	72
SOYBEANS	58
PEANUTS	42
BEANS	40

COST OF PRODUCING 100 LBS. TOTAL DIGESTIBLE NUTRIENTS

PASTURE	250 DAYS.	$.29
ALFALFA HAY	4.7 TONS.	.49
CORN SILAGE	13.7 TONS.	.91
GRASS SILAGE	13 TONS.	1.17
OATS	66 BU.	1.20
CORN	63 BU.	1.30
BARLEY	40 BU.	1.40

FOOD VALUES

PERCENTAGE OF CALCIUM·PROTEIN·FAT

PERENNIAL NATIVE GRASSES	SUMMER	.394	9.01	2.67
	WINTER	.368	3.71	2.04
	ANNUAL	.381	6.36	2.35
PALATABLE WEEDS	SUMMER	.480	9.82	2.89
	WINTER	.499	6.22	2.34
	ANNUAL	.490	8.02	2.62
SHRUBS	SUMMER	.848	12.08	7.36
	WINTER	.956	6.75	4.81
	ANNUAL	.952	9.42	6.09

8.3 SOD CROP FACTS

Another important statistic comes from the Missouri Agricultural Research Station, which found that on land sloping 3 degrees, continuously cultivated soil will erode 7 inches in twenty-four years. Grown to corn, the same land will erode 7 inches in fifty years. Grown to wheat, the 7 inches will erode in a hundred years. But sod, continuously sown to pasture, will erode only 7 inches in three thousand years. These statistics are high recommendation for sod crops.

But what sod crops should the homesteader choose? Colonial farmers knew that grass and legumes could be grown together beneficially. If grass is grown without a legume companion, the soil becomes deficient in nitrogen. The crop will contain excess carbohydrates, providing energy at the expense of its protein and mineral content. If a pasture is seeded to legumes without a grass companion, more nitrogen becomes available than is required. Volunteer grasses and weeds then invade the legume crop.

In general, it is prudent to sow a mix of equal parts of grass and legume. As mentioned earlier, legumes alone obtain nitrogen from the air. Nitrogen is fixed, or made available to plants for protein-building, by bacteria living symbiotically with plants. For this reason, legumes are less sensitive than grasses to low levels, or the absence, of nitrogen in soil.

It has been found that the presence of a legume in a sod crop increases the protein content of a companion grass. You can increase the protein content of timothy hay by growing it with alfalfa. Grass also grows at a faster rate with a legume companion. You can double a grass crop yield by including red clover. Corn grown with soybeans has twice its usual nitrogen content.

Although the major portion of a permanent sod crop usually consists of self-seeding perennials or long-lived annuals, it may also include a rapidly established, aggressive, and short-lived species, such as Italian ryegrass. There is no hard-and-fast rule for determining what constitutes an optimum seed mixture for a sod crop. At one extreme, proponents of "shotgun mixtures" advise the indiscriminate combination of up to a dozen different kinds of grass seed. Other experts advocate a mix of only two seeds, a grass and a legume. Somewhere between these two positions lies sensible practice for the small homesteader.

A well-balanced pasture community depends as much on companionable species of grasses as on climate and soil. Plants properly mixed will combine to form dense ground cover. For example, broom grass fills in around alfalfa and ladino clover fills in around orchard grass. George Washington recommended orchard grass as "the best mixture with clover."

Every homestead should have one permanent pasture. It is little wonder that advocates are so enthusiastic about this method of preventing erosion. This kind of pasture may also be companionably grown with scattered tree crops, a practice common in southern forests where longleaf pine grows in open stands with an understory of sod crops. Most sod crops grow best and produce a higher level of protein when grown in partial shade.

For some inexplicable reason, the yield of a sod crop is greater when it has undergone freezing during the dormant period. It is preferable, therefore, to broadcast seed on the ground where it will be subjected to the freezing and thawing of winter. Broadcasting seed in early winter by hand or with a hand-operated whirlwind seeder, such as the Cyclone, dispenses with expensive drilling equipment. It also avoids tractors, which compact wet springtime soils. You should remember, however, that you may need nearly twice as much hand-scattered seed to plant the area as you would if you used tractor-drawn equipment.

Mulch planting is as important to sod cropping as it is to

CYCLONE
BROAD CAST
SEED SOWER

←— HOWARD
ROTACASTER

**8.4
MULCH
PLANTING**

other farmstead crops. Nothing will better ensure a successful sod crop than applying light mulch at the time of seeding.
Mulch conserves surface moisture and delays seed germination until the soil is sufficiently warm. It also protects seedlings from excessive wind and sun and from undesirable
competition with other plants.

The properly managed sod crop requires no tillage, no artificial fertilization, and no weed-burning. You should begin a
pasture with a luxurious stand of tall weeds and brush. Then
use either a mowing machine or a stalk shredder to cut, beat,
and scatter this top growth. You can then seed in the remaining stubble, which protects seedlings as they sprout. This
heavy trash covering is highly desirable. Stubble, incidentally, offers no competition for moisture or soil nutrients
needed by young crops.

Mowing is the one management practice essential for successful sod cropping. It satisfies a number of requirements by
(1) providing hay or silage from mown clippings, (2) allowing
tough parts of the plant to decompose and improve the soil,
(3) providing uniform grazing and consequently greater total

crop production, and (4) maintaining an immature and more nutritious stage of crop growth. You should cease mowing activities in early fall so that a generous ground cover can grow for late fall and early winter grazing, extending the crop throughout the entire season. One important thing to remember when mowing or grazing sod crops is to take half and leave half. A plant's ability to manufacture foodstuff should be consistently maintained. Overgrazing or mowing too close will impair it.

To extend the grazing season to as much of a year-round program as possible, it may be necessary to have two or more separate pastures—one for cool-season and one for warm-season sod crops. Warm-season crops begin growing in late

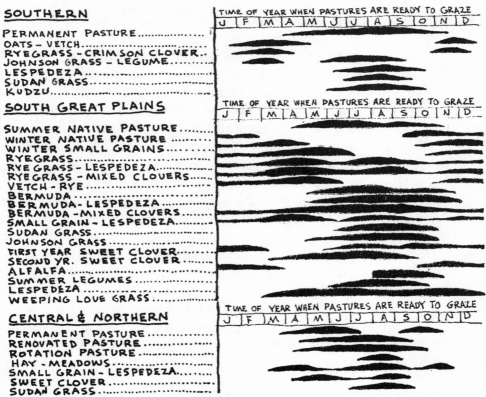

SOUTHERN

TIME OF YEAR WHEN PASTURES ARE READY TO GRAZE
J F M A M J J A S O N D

PERMANENT PASTURE
OATS – VETCH
RYEGRASS – CRIMSON CLOVER
JOHNSON GRASS – LEGUME
LESPEDEZA
SUDAN GRASS
KUDZU

SOUTH GREAT PLAINS

TIME OF YEAR WHEN PASTURES ARE READY TO GRAZE
J F M A M J J A S O N D

SUMMER NATIVE PASTURE
WINTER NATIVE PASTURE
WINTER SMALL GRAINS
RYEGRASS
RYEGRASS – LESPEDEZA
RYEGRASS – MIXED CLOVERS
VETCH – RYE
BERMUDA
BERMUDA – LESPEDEZA
BERMUDA – MIXED CLOVERS
SMALL GRAIN – LESPEDEZA
SUDAN GRASS
JOHNSON GRASS
FIRST YEAR SWEET CLOVER
SECOND YR. SWEET CLOVER
ALFALFA
SUMMER LEGUMES
LESPEDEZA
WEEPING LOVE GRASS

CENTRAL & NORTHERN

TIME OF YEAR WHEN PASTURES ARE READY TO GRAZE
J F M A M J J A S O N D

PERMANENT PASTURE
RENOVATED PASTURE
ROTATION PASTURE
HAY – MEADOWS
SMALL GRAIN – LESPEDEZA
SWEET CLOVER
SUDAN GRASS

8.5 PASTURE CALENDAR FOR YEAR-ROUND GRAZING

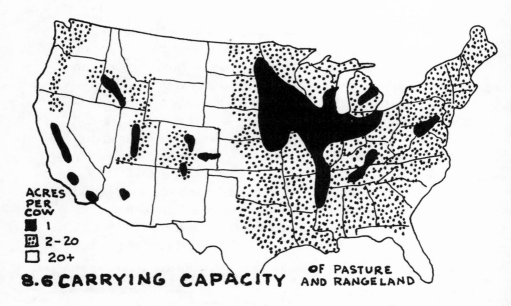

ACRES
PER
COW
■ 1
▨ 2-20
☐ 20+

8.6 CARRYING CAPACITY OF PASTURE AND RANGELAND

spring, with their primary growth achieved during summer. Cool-season crops are seeded in winter but begin their active growth only in the following spring. A dormant period occurs in the crop in midsummer, but growth resumes in early fall.

Year-round grazing, easily achieved in mild or moist climates, requires planning in regions where the ground remains frozen or snow-covered much of the winter. Since pastures are rarely productive for more than a few months at a time, two or more permanent pastures and perhaps several fields for raising supplementary feed are required to "weather through" grazing stock in northern regions. It may be advisable to incorporate ration grazing, by which a small strip of pasture is cut off from the main field each day by a movable electric fence.

The forced feeding of animals confined to dry-lot pens is an abhorrent practice. Humane considerations aside, the homesteader will discover little advantage in growing, har-

8.7 SOD CROPS

A - ANNUAL
P - PERENNIAL

L - LEGUME
G - GRASS

INTERMOUNTAIN
NORTH GREAT PLAINS

PACIFIC COAST 7

NORTHEAST 1

CENTRAL-LAKE

4 2 5

HUMID SOUTH 3

SOUTHWEST 6

VALUE:
3 - VERY HIGH
2 - HIGH
1 - AVERAGE

S - SOIL BUILDING
P - PASTURE
H - HAY

	NAME		REGION		SEED	GRAZE	SOIL	REMARKS	USE
P	ALFALFA	L	1234567	3	F-SP	SP-F	NOT WET OR ACID	DANGER OF BLOAT	H-P
P	BAHIA	L	12345	3	F-SP	SP-F	SANDY, SLOPE OK.	LOW GERM	H-P
P	BEGGARWEED	L	3	2	SP-SM	SM-F	SANDY LOAM	SCARIFIED SEED	P-S
P	BERMUDA	G	36	2	E.SP	E.F	LOAM, CLAY, SILT	SOD TRANSPLANT	P-S
P	BROME	G	12345	2	SP-F	SP-F	ANY TYPE	DROUGHT-RESIST.	HPS
P	BUCKWHEAT	G	12457	1	SP-SM	SM-F	ANY TYPE	QUICK-GROWING	H-P
P	BLUESTEM	G	45	2	E.SP	SP	ANY TYPE	DRYLAND RANGE	HPS
A	BUR CLOVER	L	37	2	L.SM	F-SP	WELL DRAINED	RESEEDS	PS
P	CARPET	G	3	1	E.SP	SP-F	MOIST-SANDY	TIGHT TURF	PS
P	CHESS	G	6	1	F	SP	LOAMS		PS
	CLOVER—								
P	ALSIKE	L	12345	3	F-SP	SP-F	WET, ACID, HEAVY O.K.	SEED FROZEN GR.	HP
P	CRIMSON	L	234567	3	F-SP	SP-F	WET, ACID, HEAVY O.K.		HP
A	CORN	G	1234567	1	SP-SM	SM-F	WIDELY ADAPTABLE	SEED WARM GR.	H
A	COWPEA	L	367	2	SP-SM	SM-F	SANDY LOAM	DROUGHT-RESIST.	H S
P	GRAMA	G	45	2	SP	SM	ANY TYPE	ALKALINE	PS
P	INDIGO	L	36	2	SP-SM	SM-F	SANDY LOAM		PS
P	JOHNSON	G	3	2	E.SP	SP-F	LOAM-CLAY	DROUGHT-RESIST.	PH
A	KALE	G	1234567	1	SM-F	SP	ANY TYPE		PH
P	KEN. BLUE G.	G	12	2	E.F	SP-F	LOAM-CLAY	LEADING PASTURE	PS
P	KUDZU	L	3	2	SP	L.F	NEEDS GOOD SOIL	GULLY, WOODS	HPS
P	LADINO	L	123	3	F	SP-F	WELL DRAINED	GRAZE LIGHTLY	P
A	LESPEDEZA	L	3	2	E.SP	SM-F	WELL DRAINED	FATTENING PAS.	HP
P	LUPINE	L	13	2	SP.F	SM-SP	SANDY LOAMS	SOUR-ACID SOILS	HP
P	MEADOW F.	G	1	2	E.F	SP-F	LOAMS-CLAY	DISAPPEARS	P
P	MILLET	G	1	1	SP-SM	SM-F	SANDY LOAM	FAST GROWTH	HP
P	MUSTARD	G	1	1	SP	SM	LOAMS		HP
P	OATS	G	1234567	1	SP.F	SM-F	ANY TYPE	WINTER GROWTH	HP
P	ORCHARD	G	123456	1	F-SP	SP-F	ANY TYPE	GROWN BUNCHES	P
P	PEAS	L	1234567	2	SP.F	SM	HEAVY LOAM	COOL WEATHER	HPS
P	RAPE	G	1247	2	SP-SM	SM-F	LOAMS		HPS
P	RYE GRASS	G	1234567	1	F.SP	SP-SM	ANY TYPE	WINTER COVER	HP
A	SORGHUM	G	126	1	SP-SM	SM-F	LIGHT LOAM	DROUGHT-RESIST.	HP
A	SUDAN GR.	G	1234567	2	SP-SM	SM-F	ANY TYPE	FAST GROWTH	HP
A	SUNFLOWER	G	1234567	2	SP-SM	SM-F	ANY TYPE	NO ACID SOIL	HP
A	SOYBEAN	L	23	2	SP	SM-F	ANY TYPE	GOOD HAY	HP
P	TIMOTHY	G	124	2	F-SP	SP-F	NOT SANDY	FAST GROWTH	PH
P	TREFOIL	L	13	3	E.SP	SP-F	WET SOIL O.K.	W/ KEN. BLUE GR.	HP
A	VETCH	L	3	2	F-SP	SP	LIGHT SOIL	DROUGHT-RESIST.	PS
P	WHEAT	G	45	2	E.SP	SP-F	ANY TYPE	EASY TO START	P

vesting, hauling, processing, storing, and rationing feed for animals suffering such imprisonment. An animal can provide itself with food from the congenial and far more sanitary surroundings of its own grazing pasture. Let the animal harvest its own food and spread its manure. The grasses animals graze from immature crops are more palatable and higher in protein and other valuable nutrients than grasses from mature crops harvested for hay or silage. Research at the Virginia Agricultural Experiment Station shows that the nutrient value of immature, uncut pasture is 50 percent greater than that of the same crop matured and cut for hay.

One final possibility for increasing sod pasturage involves grazing different combinations of animals simultaneously. Sheep and dairy cows do well in pasture together. Sheep will eat weeds and plants that cows refuse, those plants not required for high milk production. Goats tend to favor woody vegetation, thereby preventing brush from overtaking a pasture. It is good practice to let beef cattle and hogs pasture together. Hogs eat much of the vegetation that cattle overlook or waste. Figure 8.7 should provide the homesteader with some assistance in choosing sod crops for livestock food production.

9
Forage Crops

Nothing is so disadvantageous as to cultivate land in the highest style of perfection.
Pliny the Elder

No one has ever advanced a scientific reason for plowing.
Ed Faulkner, 1943

There is less need for the plow and the artificial application of fertilizer when plants, animals, *and* soil are grown together with conscious supervision that takes them all into account. The benefits and the harmony of this balanced relationship are set forth elsewhere in this book. Farmers of the Far East have been notably successful integrating crop farming with livestock-raising. Western man, however, has emerged with his notorious urge for specialization and division of labor. This appears in our agriculture as the great division between cultivators and herdsmen. The settling of the American West was rife with the struggles and problems of these two factions.

Today in the United States the struggle between commercial agricultural interests and the ecoagriculture movement may be defined, albeit simplistically, as between those who do and those who don't. Those who do are the practitioners of an agriculture that subscribes to a kind of clean cultivation and use of hard chemistry, and those who don't are ecoagriculturists who practice a gentle return of organic matter to soil . . . so nourished and protected by nature through eons of time. We covered the application of commercial fertilizer earlier. In the next chapter we discuss tilling, or cultivation *after* plowing. We will appropriately preface this discussion of forage crops with comments on plowing and its destructiveness, inasmuch as rangeland and pastureland should *never* fall under a plow.

The photograph in Figure 9.1 was taken by Richard St. Barbe Baker, leader of a 1952 Sahara expedition. It is an aerial view of a Bedouin encampment in the Libyan desert, typical of almost half of the African land mass. In better times this land was heavily forested, but it was later cleared and tilled to provide grain for Rome. As Baker observed at the time, "An iron plow is a dangerous implement, because it loosens the earth to a considerable depth, allowing the soil to be washed away in the first torrential downpour." In equatorial regions especially, the clearing of extensive areas of forest for the production of crops, such as corn and cotton, leads to certain disaster—even to the decline and fall of otherwise thriving civilizations like that of the Maya.

It is arresting to think that man has been plowing—without good reason—since the close of the Neolithic era. Early man's

9.1 BEDOUIN ENCAMPMENT.
PHOTO: R. ST. BARBE BAKER

practice of plowing and planting continues today in the Near East, where a plow stirs the soil while a farmer drops seed through an open-ended tube attached near the plow head. In this manner early farmers established a rectangular pattern of planting. (See Figure 13.1.)

Pliny the Elder's concern for the practice that continually strips away earth's soil cover has been well documented, yet with few exceptions, farmers today ruthlessly denude soil of organic matter. In their effort to create a "style of perfection," they use a heavy piece of tractor-drawn equipment called a breaking plow, which destroys weed growth, nature's most persistent protector of the soil. Plowing cuts loose and inverts slices of earth. Live, nutritive topsoil is turned under and subsoil turned up, exposing it to the beating action of rain and the scattering action of wind. Soil aggregates deteriorate with this treatment. The resulting small pores cause the denuded soil to harden and clump. This results in the need for yet more pulverizing of these clods of earth with abrasive tillage equipment. The restricted water intake of this rocklike surface crust creates the conditions for runoff and ensuing erosion, reducing the amount of moisture stored in the subsoil for use by future crops. In this manner plowing divests soil of its cover and abandons it to the elements.

In our opinion there is no suitable plowing implement. Plowing merely begins the disorganization of soil aggregate. During the first pass of a tractor over a field, 80 percent of the compaction that takes place will occur. The softer the soil surface, the more horsepower it takes to move this machine and its trailing equipment across a field. As the heavy equipment sinks into pliant soil, more and more power must be expended for the tractor to climb up out of its tracks. This is called rolling resistance, and it intensifies soil compaction. Using discs, harrows, and cultivators to rework and break up this compaction causes further compression and destruction of soil aggregate. The whole practice is then repeated the following season or with subsequent crops. Breaking the soil surface (plowing), pulverizing the aggregate (tilling), and compacting soil under heavy equipment produce slower rates of aeration and moisture infiltration, rooting problems, and reduced crop yields.

In spring, when surface moisture is right for plowing, the subsoil—always higher in moisture than topsoil—is vulnerable to maximum compaction by machinery. Soil compaction takes place throughout the whole process from plowing and cultivating to harvesting. Continual compression reduces air cavities in soil structure and limits water absorption. Soils so abused become soggy with excessive rainfall, puddling and stifling germination and root development.

Gaseous exchange must be maintained around germinating seed so that oxygen can strike the spark that activates the digestion of stored nutrients. Carbon dioxide is exchanged in this process, dissolving in large, water-permeated pores and circulating through soil by capillary action. In its form as a weak acid (carbonic acid) it dissolves minerals, particularly phosphorus and potash, rendering them assimilable by plants. It is clear that maximum pore space is essential for robust plant chemistry.

Compare your soil samples with some that are under cultivation in an average farm field. Plowing and clean cultivation have probably drastically reduced the number of large soil pores on land that has been continuously farmed. Overcropping and monoculture have taken their toll of microbe populations and nutrient stores. The repetitive propagation of one crop species makes chemical and biological demands on soil that cause it to lose its crumb structure.

A revolt against the use of the great soil destroyers—plows

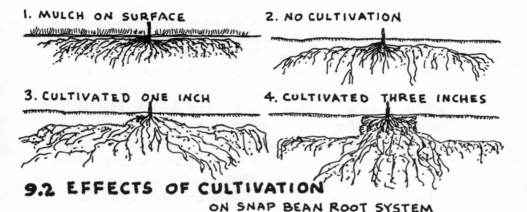

9.2 EFFECTS OF CULTIVATION
ON SNAP BEAN ROOT SYSTEM

1. MULCH ON SURFACE
2. NO CULTIVATION
3. CULTIVATED ONE INCH
4. CULTIVATED THREE INCHES

and tillage equipment—is long overdue. When Jethro Tull invented the moldboard plow, he claimed that "tillage is manure." We now know scientifically that soil tilth is created by decay, not implements. The primary reason pioneer farmers plowed land was to rid it of weeds, giving the farmer time to prepare the seed bed before wild vegetation recovered from its plowing setback.

Only recently did stored hay and silage begin to provide valuable supplemental off-season feed for livestock. In previous times livestock foraged coarse feed from rangelands or crops (in season), such as alfalfa, field corn, vetch, field peas, kale, and marigolds. By definition, rangelands are never irrigated, relying on seasonal rains; pastures are watered by either rainfall or irrigation.

Grazing takes place on rangeland, native (grass) pasture, sown permanent pastures, and leys, pastures that are sown to annual root crops. The homesteader can harvest forage that is native or planted and feed it to livestock in barn or barren pasture. You feed a cut forage crop in one of three ways: green (as a soiling crop), dry (as hay), or as silage. Whichever method of feeding you use, livestock must have a balance of dry food and succulents, roughage and concentrates, fresh and conserved foodstuffs.

You should double-crop forage; that is, you should grow crops having different requirements for planting, growing, and harvesting together in the same field or pasture. As explained earlier, roots of different species of plants will not compete with each other for soil nutrients. In this way plants with differently shaped leaves, stem structures, and growing rates will share available sunlight and soil resources.

Probably the most popular crop combination in the United States is soybeans and small grains. Winter barley double-cropped with soybeans is a particularly common practice, since barley can be harvested early enough in the spring to permit the normal development of the later-blooming soybean. Rye, barley, wheat, or oats are often double-cropped with corn to produce a silage crop. You should plant corn immediately after the harvest of small grain, to ensure that the corn has a sufficiently long growing season.

The classic example of double-cropping comes from the Aztec Indians of Mexico during the first century B.C. Corn, beans, and squash were planted successively, the corn several weeks before the beans, with each plant using soil and light to its maximum advantage. Beans climbed the cornstalks and squash covered the soil, choking back weed growth. We now know that beans supply our diets with certain amino acids that complement the amino acids found in corn. Certain of these acids by themselves are incomplete. Together, in combination with other foods that do contain their complement, they create a whole protein, capable of tissue-building. Corn also supplies carbohydrates and squash adds essential fat to the diet.

Double-cropped forage thrives best when planted in mulch.

The best use for crop residues is to leave them on the ground. Although this practice may retard early growth by cooling the soil at planting time, the slower germination is more than offset by the advantage of cooler soil. During the growing season, in periods of heat and dryness, adequate moisture is vitally necessary for seed germination. It is also necessary for any second crop grown in rotation with a first crop.

There is yet another advantage to planting forage in mulch: fields can be planted under a light blanket of organic matter, long before heavy tillage equipment could ordinarily be brought onto the field to prepare a seed bed at the end of the wet season. Well-drained soils are better, however, for any seed-planting. We have already discussed how tillage machinery compacts soil, but we should remember that planting with mulch not only requires less machinery but less power as well.

Customarily, farmers have held animals in one of two ways: in a single feeding area where they rely on being fed or in fields where they feed themselves. A third way, called deferred pasture management, works in the following manner. Provide three pastures and early in spring, when forage is at the point of "range readiness," turn the animals into pasture number 1. Here they will remove the crop almost as fast as it grows. In a month or so, before this forage crop is entirely depleted, move the animals—at the height of the growing season—to pasture number 2. Let them graze this until summer, at which time turn the animals into pasture number 3,

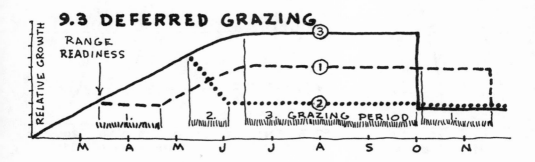

which will have been allowed to reach maximum plant growth. Let the animals graze until fall, when you should move them back to pasture number 1, where they will resume grazing the vegetation that has renewed itself in the meantime. The advantages of this program are obvious. As a result of uniform grazing, natural reseeding takes place, unwanted vegetation is controlled, and edible vegetation gains in vigor as a result of the rest period provided for each pasture.

Swiss dairymen first demonstrated the value of harvesting immature forage. Alfalfa cut in its prebloom stage has 28 percent protein and 12 percent mineral content. At full bloom it contains only 13 percent protein and 8 percent minerals. The Swiss also found that grass is more nutritious under continuous heavy grazing than when lightly grazed. Frequent cutting encourages the continuous growth of young, nutritious shoots, which are more palatable as well as more digestible. Figure 9.4 illustrates the crude protein and crude fiber content of alfalfa and Sudan grass, relative to their maturity.

J. Russel Smith is one of the trio of scientists who have im-

9.4 MATURITY AND FIBER-PROTEIN CONTENT

FULL BLOOM

BLOOMING

BUDDING

VEGETATIVE

PERCENTAGE

15 20 25 30
CRUDE FIBER

0 10 20 30
CRUDE PROTEIN

pressed us in our search for meaningful homesteading material. Working at Columbia University and specializing in tree crops, Professor Smith believed, like Professors Semple and King, in the diversification of food crops. Smith proposed using a wide variety of trees for forage. Nut- and fruit-bearing trees—such as the almond, olive, evergreen oak, honey locust, black walnut, persimmon, mulberry, plum, Chinese chestnut, and the carob—are among those that produce valuable forage and enrich the pasture. Twelve carob trees grown on half an acre will produce enough carbohydrate in one season to raise a pig from its weight at birth to 200 pounds. Six mature acacia trees on an acre will produce as much as a ton of nutritious leguminous pods for livestock consumption and, at the same time, will deposit leaves that increase the protein and mineral content of the surrounding soil.

Some well-known plants are useful as forage, too, although they are thought of only as ornamental or undesirable. Few farmers, for instance, view spineless prickly pear cactus as a foodstuff suitable for animal fodder, yet dairy cows have been known to produce high yields on rations more than half of whose content was spineless cactus. Planted in east-west rows, cactus also provides excellent wind protection for other forage plants.

Leguminous shrubs and trees planted among pasture grasses increase the grasses' nutrients. They derive nitrogen and minerals from the subsoil by the deep roots of these legumes as well as from nitrogen- and mineral-rich leaves deposited on the soil. Mulch and light shade from these shrubs and trees keep the underlying soil cooler, reducing the rate of oxidation in humus.

Where irrigation water is available, orderly cultivation of pasture crops systematically provides forage. The methods employed for such management are called deferment and rotation. Deferred and rotated crops are useful on sloping land; you should advance them downslope each year. Grow strips of row, sod, and pasture crops in sequence on level contours so that none occupies the same soil more than once every six years. For example, alternate a 20-foot-wide strip of row crop with a parallel 40-foot-wide strip of sod crop. Above the row crop should be a field of permanent pasture and below the sod crop should be an irrigated hayfield. Each year, seed a 4-foot-wide strip of sod crop to row crop. Likewise, seed

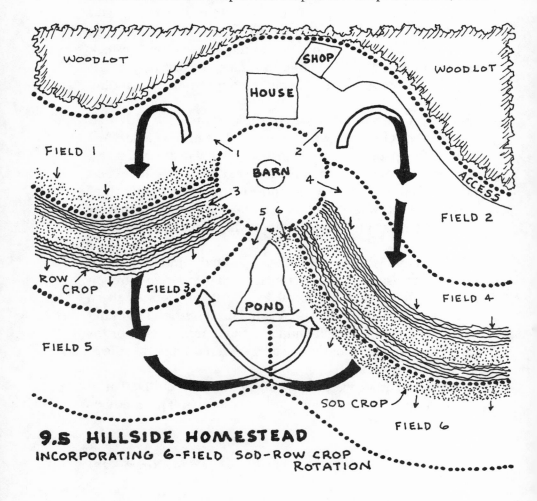

9.5 HILLSIDE HOMESTEAD
INCORPORATING 6-FIELD SOD-ROW CROP ROTATION

the upper border of the row crop to grass and legume sod crops. Intersperse forage trees throughout the whole.

The model hillside homestead, illustrated in Figure 9.5, is divided into six equal fields, each under an acre in size. Here, at all times, livestock have access to at least a single grazing field; one parcel is allocated, for example, as an irrigated hayfield. One of two fields combining sod and row crops produces silage crops. As crop areas advance downhill, part of the receding section is fallowed or hogged down during the appropriate season. A fish pond is located at the lowest point of the land, where it receives used irrigation and "clean" waste water by gravity.

With this presentation we hope you will be encouraged to integrate animal and plant husbandry. Animals utilize forage that is inedible to man, harvesting it economically. Range- and pastureland, unproductive for people's use, is thereby brought into the food chain—an inestimable accomplishment. At present, however, this is not occurring throughout the world. According to FAO, the bulk of the world's pastureland is producing less than half its potential.

10

Hay Crops

Let us never forget that cultivation of the earth is the most important labor of man. When tillage begins, other arts follow. The farmer, therefore, is the founder of human civilization.   Daniel Webster

In Webster's time the standing of tillage as an art was not in question as it is today. Commercial farmers—who now command large acreages, till land with heavy equipment, and use hard chemistry to farm crops—are finding Webster's view of agriculture ecologically and monetarily costly. To the civilization they presume to serve, their labor is aesthetically unworthy of the label of art, for these farmers have become servants of the very means and equipment they employ. They inevitably resort to plowing, which breaks up and buries topsoil to depths of several feet; cultivating, which churns and pulverizes the broken soil; and using the sprayers and heavy harvest machinery that must follow. They are addicted to the use of herbicides, fungicides, and insecticides in a vain attempt to rescue today's agriculture from its turbulent condition, yet they are inexorably losing ground. We feel compelled to reiterate that this approach to soil and agriculture will, in the end, be self-defeating. The important art of tillage, it appears, must pass back into the hands of those simple practitioners who live close to the earth. Healthy food for human civilization is truly a work of art.

In this chapter we assess the traditional methods of growing and storing hay and the ways in which it might be advantageous to modernize the process. We recommend the no-till method, which involves sowing seed in undisturbed soil, covering it with light mulch, and creating a soil climate conducive to germination and the growth of mature, healthy plants.

No-till planting prepares soil using lighter, less powerful equipment. The use of any equipment, however, raises the cost of farming, particularly of haymaking, which requires considerable time and physical effort or the use of expensive machinery for tillage and harvesting. The energy costs of the tilling and no-tilling methods are, nevertheless, quite different. Any equipment used in farming requires energy, measured in calories, for its operation. A Chinese rice farmer, using hand labor, expends 1 calorie of energy to produce 50 calories of foodstuff. Mechanized labor, on the other hand, expends 20 calories of energy for each calorie of foodstuff harvested. Chinese agriculture, therefore, is six thousand times more energy-efficient than our own. The energy requirement for conventional large-acreage tilling is 31 horsepower per hour for each acre farmed. Farmers who plant by the no-till method reduce that figure to 11 horsepower.

You may seed pasture, usually in the fall or early spring,

OPERATION	HORSEPOWER HOURS/ACRE	CONVENTIONAL EQUIP.	LABOR	TILL-PLANTED EQUIP.	LABOR
DISC	2 TO 5	$1.58	.14	—	—
PLOW	5 TO 22	3.52	.39	—	—
HARROW	2 TO 5	1.20	.14	—	—
PLANT	.5 TO 2	1.95	.21	2.64	.21
ROTARY HOE	—	.97	.10	—	—
CULTIVATE (1)	2 TO 7	1.64	.23	1.64	.14
CULTIVATE (2)	2 TO 7	1.64	.14	1.64	.14
HARVEST	3 TO 8	—	.56	—	.56
COST/ACRE		$12.50	$1.91	$5.92	$1.05

10.1 TILLAGE COSTS

either by hand broadcasting or by using a Cyclone seeder. Fall planting is preferable because a plant can establish a vigorous root system during winter months. With the arrival of spring sunlight, grass grows rapidly. Whether it is fall- or spring-planted, you usually sow a commercial hay crop on land that has first been plowed, disced, and harrowed. These conventional tillage practices are not, however, essential to a hay crop. You may accomplish sowing simply with a power-driven grain drill which evenly distributes and automatically sows to the correct depth, as described later in this chapter.

Tractor-drawn seeding equipment is advisable for homestead haying projects much over an acre in size. But buying equipment is a consideration that the homesteader must approach with prudence and caution. While a $300 Cole planter may be the most efficient stubble planter available, a $12 hand-operated Cyclone broadcaster may be entirely adequate for planting small acreage. The Cole Multiflex hugs the ground as it plants seed at uniform depth.

At several places in this book we emphasize the fact that the healthiest food for farm animals—that is, the most complete, natural food and the most economical to produce—is fresh young grass. In most areas of the United States it is not possible, however, to maintain a year-round, endless succession of high-grade pasture. Seasonal fluctuations create an overabundance of fodder in springtime and a contrasting shortage during winter. It has long been the practice of farmers to put up hay, thereby bridging the gap between periods of sufficient and insufficient fodder.

In the past, sickles, then scythes, and finally horse-drawn cutter bars were used to mow hay. Following a 1- to 3-hour period of field cure, the hay was raked into windrows, long narrow rows of cut hay about 12 feet apart extending the length of the field. Raking was done by horse-drawn equipment. The windrow stage of haymaking protected hay from rain and dew and from loss of carotene caused by exposure to sunlight. The leaching action of just one rain may reduce the nutritive value of hay by 20 to 40 percent.

Windrows were pitchforked into large cylindrical bunches, called shocks. In this form the hay continued to dry but at a much slower rate. Finally, shocks were loaded onto wagons and transported to the hayloft for final storage. Finished hay has a sweet aroma; leaves are intact on stems; and the mass is green in color and free of mold.

Haying is seemingly just a routine chore, not particularly difficult. But the beginner should be wary of the long hours and physical involvement in this task. One of the authors was once instructed in the fine art of loading the hayfork, for there is even a special technique for picking up hay. You should weave prongs of the fork in and out of the first layer of hay so that a matted foundation is formed for the remainder of the hay to rest upon. When you pitch hay onto the wagon, first place large wads along both sides to hold the central mass in place. At the end of that first day of haymaking the author, then a novitiate homesteader, came away convinced that other prospective homesteaders not instructed in the art of pitching, loading, and unloading hay into a barn loft would likely fail this task. The work would be almost sufficient to break his back.

Experienced farmers acknowledge that haymaking is the hardest, most puzzling chore of farming, and they agree that timing is critical to good results. An old saying has it that "hay cut after noon would better be cut the next day before noon." For every day's delay in harvesting a crop after June 1, the feeding value of the crop is known to decrease 1 percent.

If handled carelessly or improperly, hay can lose a third of its food value between the time of its summer mowing and its winter use as feed. If it is harvested at too late a stage of maturity, if it is allowed to dry too long in the field (sunshine bleaches out nutrients), if the humidity is too high at the time of cutting, or if the hay is mishandled, then the value of the stored feed depreciates accordingly.

It is interesting to note that leaves of pasture grass tend to dry at a more even rate than stems. However, under the most nearly ideal drying conditions, the leaves of the plant reach a moisture content for safe storage long before stems do. By the time stems finally reach this stage, leaves become too dry and brittle and then shatter easily. Unavoidably, many of these highly nutritious but shattered portions remain behind in the field when the bulk of hay is removed to a loft. Modern equipment, however, now crimps or "conditions" hay stems by crushing them so that they will dry as quickly as leaves.

One can recognize three stages of growth in the grass plant. The leaf stage, before the seed head forms, is low in carbohydrate yet high in protein and vitamins. The plant is highly digestible at this stage. In the bud stage the vegetative matter, containing fermentable carbohydrates, is considerably increased, while protein, vitamin, and mineral values decline. The flower stage finds the plant high in carbohydrate and fiber but low in protein and vitamins. The plant is then beyond the stage when it should be cut for hay.

As a grass plant matures, the protein, vitamin, and mineral contents decrease, crude fiber increases, and digestibility is reduced. In the Netherlands studies were made of the comparative cost of starch equivalents in grazing material and harvested forage. Where the production cost of a unit of starch in grazing material is given as 100 percent, the cost of a unit of starch in harvested hay is 140 percent. In silage the figure is 187 percent, and in concentrated feeds, 314 percent. Generally comparable figures are forthcoming from the

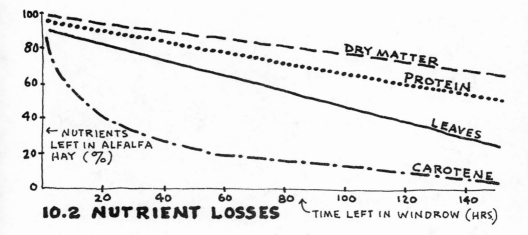

10.2 NUTRIENT LOSSES TIME LEFT IN WINDROW (HRS.)

Cornell University Agricultural Extension Service; they indicate that digestible nutrients in hay cost twice as much as digestible nutrients in grazed pasture grass. For silage the cost is three times as much as pasture, and for grain, six times.

In general, you should cut hay after the heads emerge but before they bloom. Exceptions to this are alfalfa, which you should cut at a stage of early bloom, and red clover, which you should cut at the mid- to full-bloom stage. Cut soybeans and cowpeas when the first seed pods are filled.

Volunteer hay consists of any native growth of wild grain or grass. Of the cultivated varieties, perhaps the most widely grown grains are oats, barley, vetch, clover, alfalfa, and timothy. The U.S. Census Bureau reports that in 1959 about 74 million acres of hay were cut in the United States. All but 10 million acres of this enormous crop were cultivated. It is interesting to note that 90 percent of the volunteer hay that was cut grew in the state of Nebraska.

More and more sickle-bar mowers are being replaced by rotary mowers. A sickle bar cannot condition hay so that stems dry as fast as leaves; therefore, you must operate a

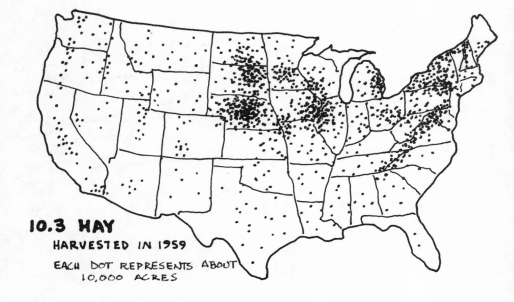

10.3 HAY
HARVESTED IN 1959
EACH DOT REPRESENTS ABOUT
10,000 ACRES

separate piece of equipment in conjunction with the cutter. Nor can windrowing be done with a sickle bar. Again, you must acquire a separate windrowing attachment or a hay rake for this operation. In dense, woody growth or when storm-damaged plants are fallen and matted, the sickle bar will not operate free of trouble.

The rotary mower, on the other hand, is a dependable, nonplugging, and highly versatile machine that will mow, cut swaths, condition, and windrow a hay crop in one pass. Furthermore, you can adjust the same machine to chop hay from a windrow. Some rotary mowers are even equipped with wagon hitches and blowers so that chopped or shredded material can immediately be blown into a pickup trailer or into a hay-baling machine.

Rotary mowers are of two types, classified according to the plane in which the cutting knife rotates: the rotary-cutter mower, having a single horizontal knife, and the flail mower, having many knives (twenty to fifty sets) that rotate in a vertical plane. In general, rotary cutters are designed for cutting and shredding stalks and underbrush, not hay. They tend to overchop hay, making it difficult to pick up.

The flail mower is ideal for homestead use in that it can mow, windrow, and condition as well as cut grass finely for silage. One machine can be made to do both jobs merely by reducing or advancing the rotating speed of knives. In addition, a flail forage harvester is equipped with a shear bar chopper, an auger, and a blower, all of which can be removed when putting up hay.

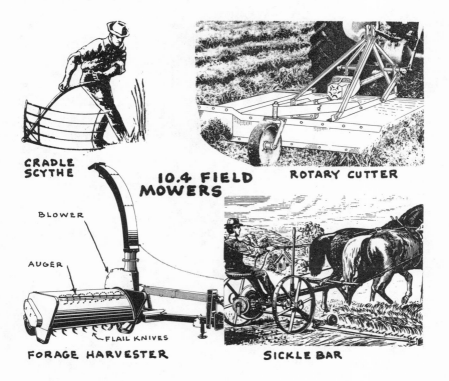

CRADLE SCYTHE

ROTARY CUTTER

10.4 FIELD MOWERS

BLOWER

AUGER

FLAIL KNIVES

FORAGE HARVESTER

SICKLE BAR

The flexible tine harrow (manufactured by Fuerst Brothers, Rhinebeck, NY) is an excellent tool for pasture management. As Figure 10.5 illustrates, the tine harrow is used to scatter manure choppings and to get rid of dead grass. It aerates the soil and stimulates new growth. The flexible harrow has a series of 3-inch-long tines linked together in a blanketlike manner. There are no rigid braces on the harrow, so it follows contours of land easily, combing the top layer of organic matter.

PASTURE RENOVATION MAXIMUM PENETRATION FOR CULTIVATION

SEED-BED PREPARATION LIGHT CULTIVATION

SMOOTHING, SCATTERING LIVESTOCK DROPPINGS

COMBINATION DEEP PENETRATION AND SMOOTHING

FLEXIBLE TINE HARROW

10.5   HAYMAKING TOOLS

CHOPPER-WAGON • UNIMOG-POWERED·SICKLE-BAR LOADER

11

Row Crops

The best fertilizer is the footsteps of the landowner. *Confucius*

The greatest fine art of the future will be the making of a comfortable living from a small piece of land. *Abraham Lincoln*

Growing food crops is an intimate affair, one stimulating and benefiting both the developing plants and their human attendants. There is, however, only one forty-centuries-old civilization, the Chinese, which to this day has a highly evolved agriculture based on row cropping cultivated by human labor.

This fact so impressed a University of Wisconsin soil scientist, F. H. King, that he traveled throughout the Orient studying Asian row-crop farming methods—particularly the Chinese. In 1910, after his return to this country, King wrote an immensely thoughtful and useful book about his travels and discoveries, *Farmers of Forty Centuries,* still a most provocative book on food production. In it King describes the plant-growing techniques that have enabled the Chinese to survive these many centuries, ever working to improve their soil structure and its fertility.

We do not advocate Chinese methods of row cropping as the ultimate in homestead food-production techniques; rather, we emphasize the sensitive awareness with which the Chinese have approached their soil and its living populations.

From the painting Going to Work, *Jean François Millet, 1848*

It is hoped that homesteaders will be inspired to an equal level of awareness.

The Chinese have long had a sense of the space in which plants grow. This space, which we call the humusphere, is that vital 6 inches of topsoil which constitutes the plant feeding zone. It is in this zone that 80 percent of the organic matter of soil is concentrated. The Chinese till their row crops very little. Instead, they broadcast legumes or cereal grains among them. When the cereals or legumes reach a growth of a few inches, the Chinese lightly work them into the soil with a hoe. Apparently, the early Chinese had an intuitive understanding of how young green manure feeds microbe populations, a point we would do well to grasp today.

We are learning through study and experience that soil tilth must be rebuilt by additions of humus-making material capable of recementing soil aggregations that have lost their crumb structure through heavy tilling. All nourishing compounds that feed growing plants are created in the humusphere by the activity of oxygen-seeking microbes. In ordinary soil conditions, the oxidation of fertilizer is a slow process, giving it time to go through an intermediate stage in which it becomes a soluble salt, ammonium nitrate. After this, salts are broken up into usable nitrogen and water for their consumption by plants. Ammonium nitrate fertilizer is an unstable, even volatile, substance that readily combines with oxygen and easily dissipates into the atmosphere. Tilling introduces that excess of oxygen which makes this likely, while at the same time it asphyxiates microbial populations.

By applying mulch as sheet compost, soil's humus layer is increased so that microbial decomposers may supply nutrients to hungry plants. Soil stability, aeration, and drainage are improved, while plant moisture and nutrition are retained. The outlay for the energy required to till is monumentally reduced. Soil so naturally fertilized and protected shows increased resistance to pathogens and pests. What commercial NPK fertilizer can do these many things?

The value of mulch as a natural manure is far greater than that of commercial fertilizer. The humus created by the decomposition of organic matter improves soil structure, and the shade it provides helps to retain moisture, minimize evaporation, increase aeration, and augment biological activity within the soil.

Admittedly, as a result of applying mulch to the surface, a delay in seed germination occurs in some climates because of the lowering of the daytime temperature of the covered soil. As Figure 11.1 illustrates, the maximum daytime soil tem-

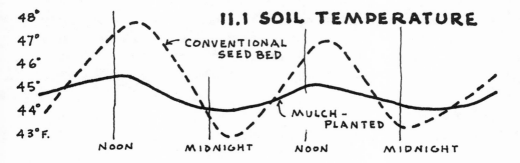

48°
47°
46°
45°
44°
43°F.

11.1 SOIL TEMPERATURE

CONVENTIONAL SEED BED

MULCH-PLANTED

NOON   MIDNIGHT   NOON   MIDNIGHT

11.2 MULCH PLANTING

1. CLEAR ROW IN MULCH 2. SET SEED ON GROUND 3. COVER WITH SAND 4. PEAT MOSS ON SURFACE 5. AS PLANT GROWS REPLACE MULCH

perature of bare ground averages 1 to 4 degrees warmer than mulch-covered soil, while the nighttime soil temperature for mulched soil is only 1 degree cooler. Thus, on the average, germination response is not appreciably decreased by sheet composting.

To promote germination and seedling growth in regions with cool soil you may, however, need to push aside several inches and plant in the cleared space, covering it with a layer of peat moss. The peat moss, being dark in color, absorbs greater atmospheric heat, keeping sand and seed warm and moist. Figure 11.2 illustrates this planting procedure. In exceptionally wet regions with heavy, clayey soil, planting in rows of sand covered with peat moss improves drainage and promotes germination and seedling development.

It would be difficult to improve on the practice of intensive row-crop planting developed by the Chinese. They also espouse intercropping, crop succession, crop rotation, and companion planting. Basically, the Chinese believe that diversification is fundamental to plant propagation. Growing various crops together permits the development of root systems of varying depth and spread—crops that feed at the same time at shallow, intermediate, and deep layers of the soil. King noted that in China as many as three different crops will occupy the same field in successive rows, all in varying stages of development. One field may contain a crop of winter wheat approaching maturity, a crop of beans about two-thirds to maturity, and a crop of newly planted cotton.

In some instances intercropping is used, and one crop is grown and harvested before the main crop requires more room. Radishes or lettuce are often intercropped with carrots, the latter being the main crop. Endive can be sown three weeks before planting a primary crop of cauliflower. An early or incidental crop may be used as a catch crop, as explained in Chapter Thirteen.

Some crops are early and some are late. An early crop is one that may be planted earlier than other crops in the regular growing season. Cabbage requires 90 days to mature, lettuce must have 50 days, and radishes need only 30 days. In climates where a 90-day growing period is sometimes risky at best, an early variety of cabbage seed might prove hardy, surviving late seasonal frost. Many combinations of early and late plantings have been worked out over the years. You may plant spring spinach with brussels sprouts, beans, or tomatoes. You may place late peas with cabbage; with warm-weather crops of either corn, beans, tomatoes, squash, or melons; or with a planting of strawberries for next year's crop.

In Chinese fields crop growth is maintained at a constant and even pace. Crops are sown successively so that they ripen as needed. When a crop is harvested, seedlings are transplanted from nursery beds to take the place of the plants that have been removed. A plant is neither allowed to stop growing nor allowed to remain in the field during its dormancy. The Chinese make good use of nursery beds (cold frames) in which winter and spring crops are started a month or so in advance of their usual season for open-ground planting. Stronger and more uniform plants can be grown in the controlled environment of a nursery bed. A suggested sequence for companion planting is illustrated in Figure 11.3.

The Chinese have a feeling for the variety of life that approaches reverence. The monotony of monoculture would re-

F.H. KING

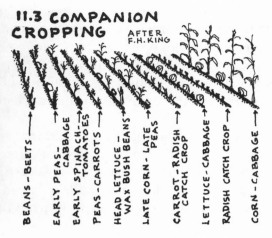

11.3 COMPANION CROPPING AFTER F.H. KING

BEANS – BEETS →

EARLY PEAS – CABBAGE →

EARLY SPINACH – TOMATOES →

PEAS – CARROTS →

HEAD LETTUCE – WAX BUSH BEANS →

LATE CORN – LATE PEAS →

CARROT – RADISH CATCH CROP →

LETTUCE – CABBAGE →

RADISH CATCH CROP →

CORN – CABBAGE →

pudiate this feeling for these people. Pure stands of plants are simply not found in nature. Monoculture does make for an efficient harvest, but disease spreads readily throughout like stands of plants, necessitating the poisonous spraying of food crops. The greater the variety of planting, the less likelihood of insect or disease infestation. Where a rich variety of plants make up a biological community, that body has a better chance to remain stable. Crop rotation and crop combination are the best ways to promote a balanced plant community. In medieval Europe, monks grew vegetables, herbs, flowers, berries, and fruit trees together for mutual benefit.

You should plan plant populations relative to the root level each species occupies in the soil and relative to the feeding capacity of each species. Follow a planting of heavy feeders, such as the cabbage family, tomatoes, and leafy vegetables, with a planting of light feeders, such as beets and carrots. An ideal program of rotation might consist of light-feeding plants, such as those from the legume family (peas or beans), which would prepare the soil for the heavy feeders to follow.

You can control bacterial blight, which characteristically attacks cabbage, by rotating cabbage with peas and beans. Tomatoes and potatoes are both attacked by the same organism, so you should never grow them in the same location every year, a practice urged, strangely enough, by biodynamic gardeners. The verticillium organism that attacks these crops remains in the soil over several seasons. Dense populations of nematodes are propagated when sugar beets are planted in the same location for several consecutive years.

The Chinese have also perfected the practice of crop rotation. They rotate shallow-rooted crops (for example, barley and corn) with deep-rooted crops (such as wheat and alfalfa). Note, as shown in Figure 11.4, that one crop may extract from the soil tremendous quantities of a particular nutrient. Rotation would alleviate this loss. The Chinese found early in the development of their agriculture that crop rotation checks disease and inhibits insect pests. As mentioned before, the fungi that feed on one particular host may linger in the soil from one crop to the next.

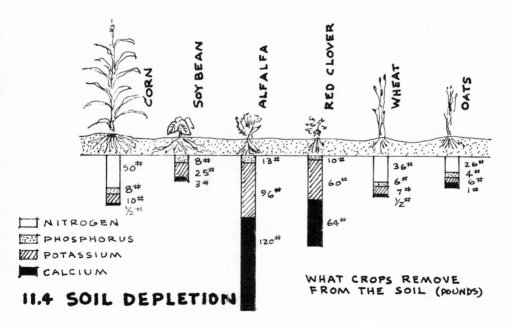

CORN · SOY BEAN · ALFALFA · RED CLOVER · WHEAT · OATS

☐ NITROGEN
▦ PHOSPHORUS
▨ POTASSIUM
■ CALCIUM

11.4 SOIL DEPLETION

WHAT CROPS REMOVE
FROM THE SOIL (POUNDS)

The double-crop system of plant rotation has been acclaimed as an effective method for discouraging crop disease and insect infestation. Generally, three crops are rotated on the same ground in a two-year period. That is, one crop of corn, wheat, or soybeans is planted once in a field over a two-year period. Debris from the crop has ample opportunity to deteriorate before the same crop is planted again. Debris of the other two crops does not harbor pathogens detrimental to the third crop.

Principles of crop rotation should be used for sod cover and animal-forage crops, as well as for row crops. During late summer dormancy a gardener would do well to encourage a healthy stand of weeds. Domesticated animals and wild creatures alike—including birds, frogs, and moles—should be allowed controlled access to garden and fields. No better method of insect control exists.

Within the expanse of pasture, woodland, orchard, and garden dwells a harmony of plants, animals, birds, bacteria, insects, and fungi. They are often considered pests, yet vain attempts to eradicate them result only in the disharmony of an otherwise balanced and diversified natural environment. In some instances insects may even prove beneficial to crops. Various strains of microorganisms, as well as plants themselves, have natural resistances and immunities to disease and pests. Pesticides tend to weaken natural crop defenses. Extermination sometimes results in an excess of fruit that should have been thinned out by normal insect activity. For example, if the moth that destroys apple buds were exterminated and all the buds on a tree allowed to bloom and set fruit, the branches would break under the weight.

Healthy soil is the best tool for biological pest control. Sir Albert Howard discovered that insect infestation and plant disease are merely indicators of incorrect farming practice.

Insects multiply out of control only by feeding on weak plants not suited to the environment in which they are growing. These plants are less than healthy, since they are unable to protect themselves against invasion. The first cause of unhealthy plants is therefore unhealthy soil. This one fact was prominent in this comment issued by the Missouri Experiment Station:

> With some of our most troublesome crop pests, there is a direct relation between insect numbers and soil fertility. The less fertility, the more insects. Our experience and studies over the last several years have proved this. In other words, as we over-crop, single-crop and permit the damage of soil erosion, we grow more crops of harmful troublesome pests than we need to have.

C. E. Yarwood, plant pathologist at the University of California, discovered that as a result of tillage, food production increased dramatically to as much as six thousand times per unit of land area—good news to agribusiness. But big yields not only reduce the nutritive content of plants; they also reduce plants' natural built-in defenses against invasion by pathogens. Yarwood states that of twenty fungal pathogens, twelve became more severe in tilled than in untilled crops. Since 1926, concurrent with phenomenal increases in crop production, the number of diseases recorded for principal crops has increased threefold. Yarwood counted 659 different plant diseases found in conventionally tilled crops and only 374 in untilled plantings.

Three factors must combine to promote the development of a disease organism. First, a virulent pathogen must exist in the immediate environment. Second, the environment must be favorable to the propagation of the pathogen. Third, a susceptible host must be present. Pathogens have the ability to survive from one growing season to another, but when land is double-cropped or planted in rotation, leftover debris from one crop may contain pathogens that, generally speaking, will not be harmful to the next crop of another kind. In the same manner certain varieties of weeds associate with specific crops. By rotating crops, nuisance weeds can be expelled. Planting with mulch of course prevents the germination of annual weeds, just as mowing before flowering time prevents the formation of weed seed.

Commercial fertilizer is also responsible for the tremendous crop yields of recent years, but concentrated applications of fertilizer are far less desirable for plant growth than a minimum balance of nutrients. Large quantities of readily assimilable nitrogen actually promote wilt disease in some crops by providing better nourishment for the parasite that causes it.

Perhaps in a few years the idea of zero tillage will have wider acceptance in agricultural practice. Major advances in machinery design and in mulch-planting techniques are now coming from England and Germany. At the Agricultural University of the Netherlands a special department has been formed to develop tillage machinery that does not compact soil. At the same time other scientists in other departments at the university are developing herbicides. The Dutch reason that zero tillage and the use of herbicides will eventually replace the need for plowing and cultivating by eliminating competitive weed growth entirely.

Most of the significant progress in zero tillage has been made during wartime, when energy resources are preempted by the military and when demands for nourishing food for armies are high. The Spanish-American War no doubt created a demand for efficient crop production and fanned

King's interest in minimum tillage. At the time of World War I Professor Hollack in Germany wrote his theory of no-plow technique, just as Edward H. Faulkner did in 1943 in his book *Plowman's Folly*. During the early years of World War II, Rothamsted Agricultural Station in England conducted three years of research on minimum tillage. At the same time in Vienna, Viktor Schauberger stated his case against the high speed with which a plow slices through soil, claiming that this causes electrical disturbances destructive of essential trace minerals. There is, however, a modern peacetime proponent of zero tillage. She is Ruth Stout, gardener extraordinaire. Like Faulkner's, none of her books are recognized by the scientific community, possibly because of her popular approach. She has, nevertheless, popularized mulch planting among organic gardeners. Hers is an effective, if not scientifically authorized, position against tilling, fertilizing, and manuring.

Rudolf Steiner, founder of the biodynamic approach to gardening, maintained that permanent, raised garden beds actually increase growth activity at the surface of these beds. There are obvious practical advantages for raised, modular gardening beds. Cold air will drain from around plants. Flats and screened frames, built to fit the module exactly, can be moved for use on other beds as necessary. This consolidated arrangement promotes the maximum use of planting space—the first principle of intensive gardening. We have found a 4-foot-wide module with a bed length of 16 feet to be a practical size.

You should consider soil fertility when you first choose a garden location. Therefore, plant where weeds have flourished. Protect the location from wind and frost. It should offer the best slope, drainage, and exposure to sunshine. Some plants, like tomatoes, will thrive in half the sunshine

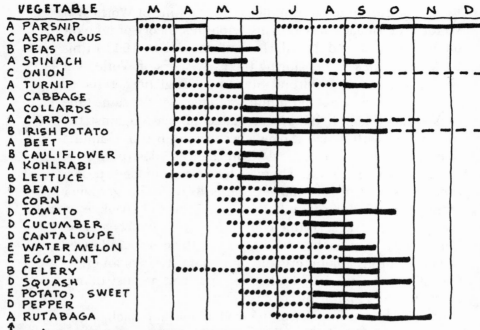

VEGETABLE	M	A	M	J	J	A	S	O	N	D
A PARSNIP										
C ASPARAGUS										
B PEAS										
A SPINACH										
C ONION										
A TURNIP										
A CABBAGE										
A COLLARDS										
A CARROT										
B IRISH POTATO										
A BEET										
B CAULIFLOWER										
A KOHLRABI										
B LETTUCE										
D BEAN										
D CORN										
D TOMATO										
D CUCUMBER										
D CANTALOUPE										
E WATERMELON										
E EGGPLANT										
B CELERY										
D SQUASH										
E POTATO, SWEET										
D PEPPER										
A RUTABAGA										

A. COOL-SEASON CROPS - SLIGHT FREEZING TOLERANCE
B. COOL-SEASON - DAMAGE NEAR HARVEST BY FREEZE
C. COOL-SEASON - THRIVE OVER WIDE TEMPERATURE RANGE
D. WARM-SEASON CROPS - INTOLERANT OF FROST
E. WARM-SEASON - REQUIRE CONTINUOUS WARM WEATHER

11.5 PLANTING & PICKING DATES

considered normal for other crops. The total, or "dead," shade of a building does not, however, encourage plant growth as the moving shade of a tree does.

Situate row crops in a place immediately accessible to the homestead cooking area. Experience has shown this to be of practical importance, conducive to the daily maintenance and harvesting required by a garden.

There are hundreds of published plans that purport to show the optimum garden layout. Actually, it is impossible to

RATING	VEGETABLE	YIELD/ACRE 100 POUNDS	MAN-HOURS PER ACRE	AV. DAYS TO PRODUCE	PROTEIN LBS/ACRE	CALORIES 1000 LBS/ACRE	VITAMIN B GRAMS	VITAMIN C GRAMS	VITAMIN G GRAMS	VITAMIN A UNITS
1	SPINACH	117	117	50	221	1058	5.5	1443	15.3	873
2	CARROT	194	243	80	210	3587	7.9	315	7.2	246
3	ONION	198	239	110	261	4103	5.3	765	3.8	0
4	SQUASH, WIN.	170	58	120	189	2516	2.7	176	4.6	171
5	POTATO, IRISH	152	130	120	255	4909	7.5	638	3.5	2
6	CELERY	320	300	120	263	2019	3.2	646	3.9	3
7	CABBAGE	137	111	90	140	1298	4.8	3180	4.5	3
8	TOMATO	110	108	180	108	1130	4.5	1119	2.4	42
9	BEAN, SNAP	46	133	55	99	787	1.4	282	2.0	23
10	LETTUCE	91	109	80	75	531	2.5	400	4.8	11
11	TURNIP	120	165	60	115	1618	3.8	1190	3.6	1
12	BROCCOLI	73	176	150	114	587	1.4	1411	5.5	94
13	CAULIFLOWER	108	151	75	148	865	4.6	1990	5.2	1
14	BELL PEPPER	69	144	150	70	786	.7	3171	1.0	132
15	POTATO, SWEET	60	127	175	94	2940	2.7	266	2.0	59
16	CORN	62	80	90	87	1154	1.4	106	.6	6
17	SQUASH, SUM.	97	146	60	57	804	2.0	132	3.5	129
18	BEET	108	218	70	97	1256	2.2	146	3.5	4
19	CANTALOUPE	98	195	160	35	734	1.3	798	2.0	37
20	ASPARAGUS	44	188	365	73	400	2.5	423	1.5	8
21	PEA	22	143	85	72	476	1.9	101	1.3	6
22	CUCUMBER	84	190	100	41	383	2.4	212	4.0	1
23	RADISH	120	273	30	70	588	2.0	429	.8	0
24	WATERMELON	103	110	160	24	670	.7	153	.7	2
25	BEAN, LIMA	14	140	80	41	324	.7	62	.6	1

11.6 VEGETABLE EFFICIENCY RATING

design a plan for general use. Site conditions, regional climate variations, and personal taste are all too variable. A few suggestions, however, may prove useful. First, as the old adage advises, "never plow up more space than your wife can take care of." In the contemporary vernacular, this means that you should keep your garden area small enough to provide your food needs while using to best advantage the space, time, and help that are available. Start small and grow with your garden.

Second, grow plants from seed of the finest quality. Every homestead family should have the experience of sitting around the fire on a stormy morning late in winter, consulting last year's carefully selected and saved seed supply or the seed catalog just arrived in the mail, preparing the rating chart for the approaching season of growth. Each crop should be considered for its growing season, its adaptability to your garden site, nutritional value, taste, and the efficiency with which it may be grown.

The vegetable efficiency chart, Figure 11.6, was originally prepared to assist us in planning and planting our homestead garden. We prepared the chart with reference to the number 6 zone of plant hardiness shown in Figure 11.7. In zones where the average annual minimum temperature is lower or higher, the efficiency rating changes accordingly.

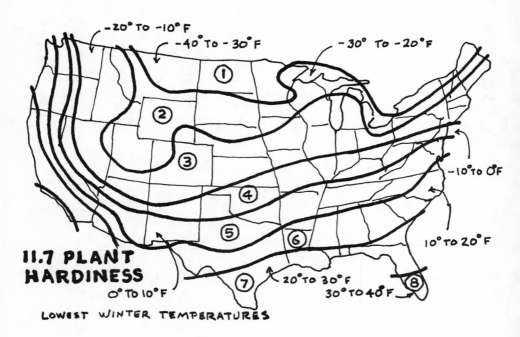

11.7 PLANT HARDINESS
0° TO 10°F
LOWEST WINTER TEMPERATURES

12
Tree Crops

Trees full of soft foliage; blossoms fresh with spring beauty; and, finally—fruit—rich, bloom-dusted, melting and luscious—such are the treasures of the orchard and the garden, temptingly offered to every landholder in this bright and sunny, though temperate climate.   A. J. Downing

At considerable cost in time and energy, many orchardists have amassed abundant knowledge supporting the concept of diversified tree cropping over the years. This knowledge, unfortunately, reaches but an occasional homesteader. It surely finds little acceptance from contemporary farmers and orchardists, most of whom repudiate change in the current practice of monocultured tree crops. A. J. Downing's *Fruits and Fruit Trees of America,* published in 1845, remains the basic reference for tree crops even today, although it is shamefully ignored.

Downing adapted western European fruit-growing practice to the requirements of the American climate. Fruit trees planted in Massachusetts and Michigan at the height of Downing's influence (1870–90) still stand and bear fruit. Thousands of trees carelessly planted in 1890–1920 have since broken and died. There is a refreshing simplicity to Downing's basic technique for tree care:

A judicious pruning to modify the form of our standard trees is nearly all that is required in ordinary practice. Every fruit tree, grown in the open orchard or garden as a common standard, should be allowed to take its natural form, the whole efforts of the pruner going no further than to take out all weak and crowded branches.

But today Downing's ideas are either little known or neglected because they do not conform to current practices of tree-crop producers in the United States.

An even more tragic example of knowledge gone abegging exists in the work of J. Russel Smith, treeman par excellence and professor of economic geography at Columbia University. Smith launched his study of trees considered commercially useless in 1910 with a worldwide quest for new tree varieties. In 1929 he published *Tree Crops—A Permanent Agriculture*. He reinforced his valuable discoveries with further world travel and in 1954 with a revised edition of his book.

As a tree devotee, Smith criticized the cultivation of annual row crops. He felt that crops that must build themselves from scratch for every harvest are but victims of the climatic uncertainty of short seasons. Tree crops, on the other hand, are not affected to the same degree by such climatic uncertainties as drought. Their deep roots enable them to accumulate water from depths unreached by most row crops.

Smith knew that the American practice of living high on the food chain was dangerous as well as wasteful. That animals consumed 80 percent of America's harvest alarmed him. He made a good case for switching to a diet of tree crops, since the proverbial meat-and-potatoes ration yields 800 calories, while an equivalent amount of nuts contains 3,200 calories. He contended that if animals were to be raised for their products or their meat they should be husbanded in a manner that would encourage them to harvest their own food. Trees, he said, were the answer, since the mulberry, persimmon, oak, chestnut, honey locust, and carob all produce excellent animal food. Today this practice of self-harvesting, known as hogging down, is considered a major innovation.

John Gifford was one who saw the social and economic implications of tree crops. In his book on diversified tree-crop farming for the tropical homestead, he demonstrated the unsuitability of annual crops in tropical climates, where deep-rooted trees thrive. The Maya failed to survive because they cut down their forests and planted corn. For the most part, treemen consider corn the killer of continents, one of the worst enemies of the human future. Annuals, they claim, are the food crop of primitive man—food needed in a hurry. Promethean man supposedly has, to Gifford's way of thinking, the culture, leisure, and intelligence to subsist on tree crops alone.

Treemen qualified in the science of pomology (fruit-tree cultivation) unanimously agree that the interplanting of various tree species is a necessary and desirable practice. Interplanting makes sense for purely economic reasons. When crops such as peaches, pears, and plums are interplanted with apples in an orchard, revenues from their yield can subsidize production of the main crop. Crops that mature rapidly, such as the dwarfed varieties, can be planted alternately with more slowly maturing species. Mulberries are an excellent choice for interplantation with nut trees. They grow rapidly, bear young, and are resistant to shade.

One type of interplanting is known as two-story agriculture. By this method trees are grown on land that is planted to other crops as well. As we stated previously, sod crops and tree crops can be mutually beneficial when planted companionably. Ten black walnut trees in an acre of permanent pasture illustrate this point. The deep roots and thin, open foliage characteristic of walnut trees can share the available space and nutrients with a lower-story sod crop. Walnuts, which grow primarily in late spring, give grass crops more time to establish themselves before the warm season. Leaf-filtered sunlight assists the continued growth of a sod crop through summer months.

There are numerous advantages to two-story planting of a fruit with a vegetable crop. Trees bear their fruit in the upper story, while brambles, grapes, bush fruit, or vegetables grow below. Long-lived fruit trees continue to bear after short-lived lower-story plants have been harvested. Placing filler trees between standard varieties provides early bearing for a fruit crop that is otherwise short-lived. Dwarfed trees function well as fillers in orchards of standard bearing trees. They can even be early-bearing varieties of the same kind of fruit. For example, set a wealthy apple between Northern Spy or McIntosh apple trees.

Planting in a pattern of squares wastes space. However, when a tree is also planted in the center of each square, creating a triangular arrangement, many more trees may occupy an allotted space, as shown in Figure 12.1.

A gardening program for an intensively planted garden plot might include the following sequence: vegetables should be harvested and ever-bearing strawberries planted the first season; a partial crop of brambleberries should be gathered the second to tenth years, after which they should be removed; dwarfed fruit trees bear from the third to the tenth year, and you should eventually remove them to allow room for the growth of semipermanent trees; peaches and plums will produce until their fifteenth year, at which time you may remove them to provide maximum space for the further development of appropriate sod crops and two remaining apples and a pear, which will continue to bear indefinitely.

There are a number of conditions influencing your choice

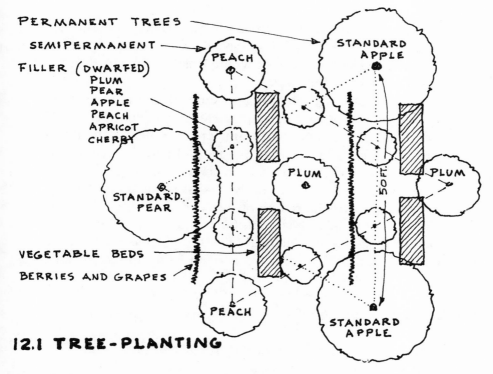

PERMANENT TREES

SEMIPERMANENT

FILLER (DWARFED)
PLUM
PEAR
APPLE
PEACH
APRICOT
CHERRY

PEACH

STANDARD
APPLE

STANDARD
PEAR

PLUM

PLUM

50 FT.

VEGETABLE BEDS

BERRIES AND GRAPES

PEACH

STANDARD
APPLE

12.1 TREE-PLANTING

of a lower-story sod crop grown with an orchard crop. The sod crop ideal for this combination is one that grows slowly at first while accompanying trees need ground moisture. It will then grow more rapidly later in the season when trees require less water. Soybeans and cowpeas grow well in this manner. Alfalfa and small grains are poor choices, however, because their extensive root systems tend to rob a tree of moisture. Leguminous sod crops, such as hairy vetch, are especially valuable for maintaining soil fertility in an orchard of nut trees.

Tree crops planted in heavy, poorly drained soil will benefit from a lower-story crop of permanent pasture, such as bluegrass or orchard grass. Roots of grass crops siphon off excess water and force an increase in the size and capacity of the soil pores and fissures needed for air. Heavy, wet soils are aerated in this way, and if you avoid tillage and compaction, surface roots will receive air where it is best obtained—at or near the surface of the soil.

Plowing, discing, rototilling, or cultivating in any manner around trees is a destructive practice from which you should strictly refrain. A homesteader must realize that the great majority of a tree's feeder roots are located within 1 foot of the soil's surface. This is the zone where soil is the most fertile and where aeration primarily takes place. Mulching is preferable for tree enrichment. Besides conserving moisture and providing additional nutrients, mulching protects tree roots from cold-season damage. Tests run in one season at the University of Kansas showed that on bare ground freezing cold penetrated to a depth of 26 inches. A cover of snow reduced this depth to 12 inches. Straw mulch combined with an additional cover of snow further reduced freeze depth to only 6 inches.

Mulch cover tends to retard spring blossoming. This is desirable where late frosts threaten, as they do in eastern Oregon, where it may freeze at any time during the growing season. Another practice that helps prevent premature blossoming is planting on a northern slope. Sun in winter and early spring does not strike a northern exposure with the same intensity as it does a southern one. Sap rises if the temperature of the tree trunk appreciably increases. Therefore, cold evening temperatures may kill newly swelled fruit buds. On slopes facing north, however, blossoming will be retarded until seasonal temperatures are well advanced. Some tree crops, like apricot, sweet cherry, and almond, blossom early. Some apple varieties, like Rome Beauty and Northern Spy, blossom late. As a rule, the blossoming of fruit begins early in the southland and moves northward at a rate of 1 degree of latitude about every five days. Altitude, of course, somewhat influences these figures. You should not adhere strictly to charts giving the "average date of the last killing frost." *Average,* in this instance, means that 50 percent of the frost indicated for a specified date occurs before that date and 50 percent occurs afterward.

Injurious winter temperatures can also be influenced by the tempering effect of situating water strategically near the orchard. There is, for example, more danger of freeze damage to tree crops in the Mississippi Valley at latitude 38 degrees than there is in ocean-surrounded Nova Scotia at latitude 45 degrees. An orchard planted on the leeside of a southern or eastern shore gains significant relief from temperature extremes. This assumes, of course, that the shore water remains unfrozen. Frozen water offers no protection whatsoever. Water-soaked soil does, however, supply a modicum of latent heat on a frosty night.

An ideal site for a tree crop is one that is higher than surrounding ground. Trees planted in a natural draw are subjected to cold-air drainage. Even in a draw with a gradual slope, cold air may be caught by the obstacle of a tree itself, engulfing and damaging it. You may enhance air drainage with careful planting, as Figure 12.2 illustrates. Trees planted on contour or in rows across a slope may impede the drainage of air.

When you are locating homestead tree crops, you should give thought to protection from wind. Winds are often accompanied by ground-saturating rains, which makes trees vulnerable to toppling. Chapter Thirteen gives information on planning windbreaks.

There are various reasons for choosing dwarfed trees in preference to standard-size trees. They circumvent some of the problems associated with site and climate. The ground-

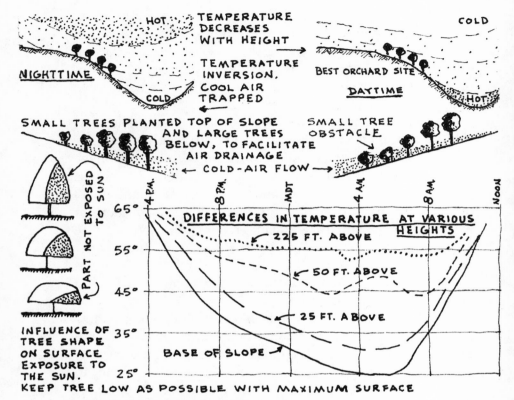

TEMPERATURE DECREASES WITH HEIGHT

TEMPERATURE INVERSION. COOL AIR TRAPPED

HOT

COLD

NIGHTTIME

COLD

BEST ORCHARD SITE

DAYTIME

HOT

SMALL TREES PLANTED TOP OF SLOPE AND LARGE TREES BELOW, TO FACILITATE AIR DRAINAGE

SMALL TREE OBSTACLE

← COLD-AIR FLOW →

DIFFERENCES IN TEMPERATURE AT VARIOUS HEIGHTS

225 FT. ABOVE

50 FT. ABOVE

25 FT. ABOVE

BASE OF SLOPE →

PART NOT EXPOSED TO SUN

INFLUENCE OF TREE SHAPE ON SURFACE EXPOSURE TO THE SUN.

KEEP TREE LOW AS POSSIBLE WITH MAXIMUM SURFACE

12.2 SITE INFLUENCES

hugging characteristic of dwarfed development permits dwarfed trees to receive more of the warmth radiated upward from the ground. There is, for example, a dwarfed apple, known as a creeper, that grows as far north as Siberia. Having little height, it absorbs any available ground heat and better withstands cold, since in wintertime it is protectively covered with snow.

Factors other than site and climate affect your choice of trees for the homestead orchard. Hardiness is one. A tree that is hardy, or vigorous, is relatively more disease-resistant than less hardy varieties. For example, in California English walnuts are grafted onto black-walnut root stock, because long years of experience showed that black-walnut roots are less susceptible to fungus attack and more readily survive California's temperature fluctuations.

Root-stock grafting is also done to attain a deeper root system in areas where rainfall is limited or where supplies of irrigation water are insufficient. Choosing the proper tree is essential in moisture-scarce regions. Cherries, grapes, and olives require less moisture than oranges, apples, and pears. In Tunisia olive trees are spaced as far as 100 feet apart to gain a moisture advantage in the arid clime.

Commercial orchards planted to the monoculture of a single tree variety are invariably overcrowded, overtilled, overfertilized, overpruned, and unnecessarily diseased. But, to be feasible commercially, a money-motivated orchard crop could hardly be operated otherwise. Tree-crop food production is one activity in which the homesteader has a major economic advantage over the commercial orchardist. The homesteader can indulge all of the practices considered uneconomical, such as intercropping with lower-story food crops planted in mulch, to produce better-tasting, more nutritious, disease-free fruits and nuts.

To remain competitive, commercial orchardists must stimulate early, large, and colorful crops. Of the numerous shot-in-the-arm techniques for attaining these ends, fertilization is probably the most widely used—and misused. Charles Darwin was one of the first scientists to point to the dangers in fertilizer application. In his book *Variations of Animals and Plants under Domestication* he refers to a statement by Gartner that warns that because of heavy concentrations of fertilizer, sterility is especially common in such crops as cereals, cabbage, peas, and beans. This surely applies to trees. Concentrations of salts found in both farmyard manure and commercial fertilizer destroy the tender feeder roots of trees. Many newly planted trees die because the grower, eager to give the trees a good start, forces massive amounts of fertilizer into root zones. A mixture of damp peat moss and loamy soil spread around roots is far better than fertilizer in any form.

Do not saturate the hole in which a tree is to be planted with water. Thoroughly moisten the surface of the soil after the tree is planted to facilitate tamping, but avoid excessive amounts of water, which provoke the formation of clods in the subsoil.

Fall is considered the best time for planting trees. In climates where severe winter weather prevails, however, spring is preferred. Some root growth will take place throughout winter months in a fall-planted tree if it is planted with heavy mulch about its roots. Direct the trunk of the tree slightly into the prevailing wind. This encourages the roots to develop to the windward side of the tree. The most vigorous roots always lie directly beneath correspondingly vigorous branches. You should grow dioecious trees, which have reproductive organs of only one sex, as you would raise a herd of animals. Plant one male (staminate) tree with a cluster of female (pistillate) trees.

When you have found a tree that has especially desirable qualities, you will want to multiply its progeny. There are two ways to do this: by raising trees from seed or by grafting branches of the desired tree onto another tree. Smith describes a good method of raising seedlings. Nail together four plaster laths—one of which has been soaked in a solution of nitrate of soda—to form a 1-inch-square tube 3 feet long. Firmly pack the tube with humus-rich soil and place the seed near the surface at one end of the tube. As the plant grows, roots will cling to the lath that was previously soaked in the solution. You can then plant the 4-inch-high seedling (which will have 3 feet of root development) deep into the ground using nothing more than a crowbar to prepare the hole. Before germination is possible, however, some tree seeds—especially nuts—must undergo some exposure to freezing.

The tree raised from seed is not apt to develop identically to the one from which it originated. For this reason the technique of grafting was developed. A small shoot or bud of a tree is inserted into another tree, where it continues to grow and become a permanent part. Grafting is, in effect, the healing in common of two wounds. Commercial nurseries charge high prices for grafted stock, and the public bears the cost of the mystique associated with tree grafting. Actually, grafting is a simple basic skill that can be mastered by anyone who will take the time and trouble to understand a few principles and techniques.

Graft a scion—which is a branch from a tree producing a desired characteristic—onto the parent tree, called the rootstock. Both the scion and the rootstock should have approximately the same diameter, about ½ inch. Graft early in spring so that the wound will heal before tree growth resumes.

Budding and layering are other grafting methods of tree propagation. The homesteader will have more success if grafting is done on trees and seedlings that are growing in their permanent locations. Figure 12.3 illustrates the technique of budding. The layering process involves inducing a part of the parent plant to grow roots, or shoots, which are then separated from the parent host for their own independent growth. You can start dwarfed fruit trees in this manner.

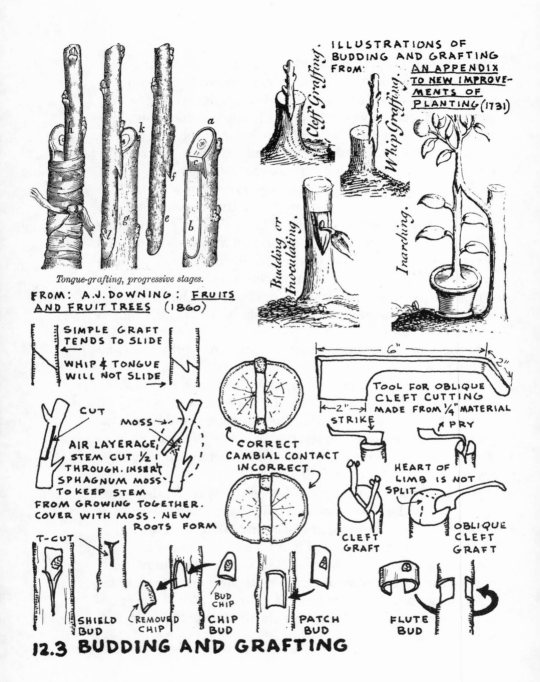

Tongue-grafting, progressive stages.

FROM: A.J. DOWNING: FRUITS AND FRUIT TREES (1860)

ILLUSTRATIONS OF BUDDING AND GRAFTING FROM: AN APPENDIX TO NEW IMPROVEMENTS OF PLANTING (1731)

Clef Grafting.

Whip Grafting.

Budding or Inoculating.

Inarching.

SIMPLE GRAFT TENDS TO SLIDE

WHIP & TONGUE WILL NOT SLIDE

CUT

MOSS

AIR LAYERAGE, STEM CUT ½ THROUGH. INSERT SPHAGNUM MOSS TO KEEP STEM FROM GROWING TOGETHER. COVER WITH MOSS. NEW ROOTS FORM

CORRECT CAMBIAL CONTACT INCORRECT

6" 2"

TOOL FOR OBLIQUE CLEFT CUTTING MADE FROM ¼" MATERIAL

2" STRIKE PRY

HEART OF LIMB IS NOT SPLIT

CLEFT GRAFT

OBLIQUE CLEFT GRAFT

T-CUT

SHIELD BUD

REMOVED CHIP

BUD CHIP

CHIP BUD

PATCH BUD

FLUTE BUD

12.3 BUDDING AND GRAFTING

A dwarfed tree is nothing more than a vigorous scion grafted onto rootstock displaying less vigorous growth. Quince rootstock takes little nourishment from the soil and requires only nominal amounts of carbohydrates for its growth, whereas the pear scion will manufacture considerable carbohydrate for its growth at the cost of its assimilation of protein. Quince rootstock will therefore dwarf the growth of a pear scion, but that pear scion will, nevertheless, tend to grow larger and faster than the original quince. The high proportion of carbohydrate over protein results in dwarfing and early fruition of this grafted tree. The northern dwarf apple bears fruit in four years, in contrast to the fifteen years' growth required for fruit from standard apples. Apple, pear, cherry, peach, plum, and apricot trees can all successfully be dwarfed and have their place in the homestead orchard-garden.

As mentioned earlier in this chapter, the dwarfed creeper apple tree, because of its low profile, is widely grown in the Siberian weather zone. In particular where conditions necessitate maximum exposure to sun and strong resistance to wind, this low, spreading tree form should be used. The shade from such a low-growing tree demonstrably conserves ground moisture. Some tree experts claim that this type of tree requires low soil fertility.

The greatest vigor in a tree is found near its crown, so by pruning its tops, vigor spreads to other parts of the tree. It has been said of pruning that there is no horticultural practice "concerning which there is a greater diversity of opinion or in the application of which there is a greater diversity of procedure." According to Bulletin 376 of the Illinois Experiment Station, the unwarranted death of trees is often mistakenly attributed to "mishandling" rather than correctly attributed to severe or badly executed pruning.

The main purpose of pruning is to remove injured and

diseased growth from a tree. Without the protective outer cover that bark offers, dead limbs are attractive to parasites and saprophytic fungi, and they should therefore be removed each season. Another reason for pruning is to train a young tree structurally, shaping it to resist damage by wind, snow, and ice at later, more mature stages of development. The central-leader pattern and the modified-leader pattern for pruning are in common usage along with the somewhat less popular open-center pattern. A tree pruned to an open center is structurally defective but does receive more interior sunlight.

As trees mature, the purpose of pruning them changes. On a mature tree limit pruning activity to maintaining a balance between the growth of vegetation and the growth of fruit. For the most part, thinning top growth maintains vertical balance between the growth of the root system and the system of vegetation. When transplanting, remove top growth to balance the remaining root system. Downing said that even late spring pruning is preferable to winter pruning, since wounds heal more rapidly while the tree is active. The old adage that advises pruning during the dormant period "when the sap is down" appears foolish in light of the fact that wood is just as sap-laden in winter as it is in spring. Moisture circulation in trees is upward from the soil, through roots and sapwood to leaves, where mineral-saturated water is converted into starches and sugars, which then travel downward through the inner bark, feeding the tree en route.

With clear advice and conscientious practice, grafting and pruning can be some of the most rewarding tasks performed on the homestead. The art of food-crop propagation may never be fully learned or mastered, but from the beginning of an association with plants, you will receive immense satisfaction. You may even attempt superlatively trained tree forms,

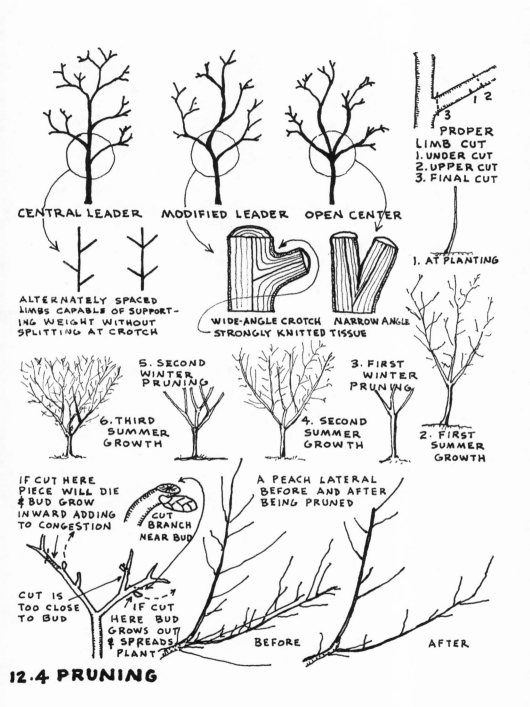

CENTRAL LEADER   MODIFIED LEADER   OPEN CENTER

ALTERNATELY SPACED
LIMBS CAPABLE OF SUPPORT-
ING WEIGHT WITHOUT
SPLITTING AT CROTCH

WIDE-ANGLE CROTCH   NARROW ANGLE
STRONGLY KNITTED TISSUE

PROPER
LIMB CUT
1. UNDER CUT
2. UPPER CUT
3. FINAL CUT

1. AT PLANTING

5. SECOND
WINTER
PRUNING

6. THIRD
SUMMER
GROWTH

3. FIRST
WINTER
PRUNING

4. SECOND
SUMMER
GROWTH

2. FIRST
SUMMER
GROWTH

IF CUT HERE
PIECE WILL DIE
& BUD GROW
INWARD ADDING
TO CONGESTION

CUT
BRANCH
NEAR BUD

A PEACH LATERAL
BEFORE AND AFTER
BEING PRUNED

CUT IS
TOO CLOSE
TO BUD

IF CUT
HERE BUD
GROWS OUT
& SPREADS
PLANT

BEFORE

AFTER

12.4 PRUNING

like espalier, but not without great skill and patience. The serious homestead propagationist can create and display exotic conversation pieces, such as a five-variety apple tree or a tomato-producing potato plant.

The biggest challenge of raising trees for food remains the development of original, unaltered tree stock. Russel Smith traveled the world searching for pristine tree stock. Today, nurseries deal only in budded or grafted stock. Seedlings can be cross-pollinated to produce even more desirable strains of trees. Orchards of exceptionally productive, true seedling stock have been extensively planted in such places as Chile and Spain. There exists the world over no more exciting or worthwhile project for homestead development than tree propagation.

13
Woodlands

If by means of forest farming world production of foodstuffs and raw materials can be increased substantially and, where appropriate, tree crops linked with industrial development, something of real significance will have been achieved—both for the better sustenance of mankind and for the preservation and enhancement of our environment.

J. Sholto Douglas

Oliver Wendell Holmes once said, "Wood, like knowledge, should be well seasoned." We believe this is far from the case with current woodland management in this country, for like agriculture management, little of its knowledge of forest and woodland use is seasoned. Present-day loggers harvest wood crops at "ground line"; that is, profit is secured from trees cut at ground level in a single harvest, with the intent of taking one kind of tree of one size for one purpose. In this way whole forests are "clear cut"—efficiently stripped of select timber. Trees too young to be commercially useful are left lying on barren ground. There is little or no selective pruning, sorting, sizing, and planning for multiple use of these wood resources. Nor is there any attempt to use the tops of those trees that are deemed fit to be hauled away. Profit-motivated forestry management uses other practices that further compound the destruction and waste of forests—crop-dusting and controlled burning.

Bad forestry poses a greater threat to a nation's welfare than bad farming, since there is more to lose. When a forest is gone it cannot promptly be replaced by the next crop. During

13.1 GROUND-LINE FARMING USSCS PHOTO

the Depression of the 1930s, the Soil Conservation Service photographed one result of ground-line harvesting. Figure 13.1 is an early photo showing the intersecting pattern made by plow runs after deforestation. Note that topsoil is completely gone. Protection by windbreaks and planting with mulch would have saved this land.

Experiments show that soil losses are a hundred times greater from deforested land than from timbered land. And water losses are sixty times greater. Our planet presently has twice as much forest area as land under cultivation, with one-third of the continental surfaces classified as forest soil. Before the white man came, America had well over half of its 2 billion acres in forest growth. In places like England, however, forests have been so badly decimated for so many centuries that forests are now simply known as "woods."

Today, woods are found on land classed as nonagricultural because of its inaccessibility, steepness, or infertility. Farmers refrain from planting forest trees or shrubs on ground that, for their purposes, is better used for quick-profit annual food crops. Food production is, after all, America's chief industry and exportable commodity. The homesteader, on the other hand, need not be restricted by a system of land use that seeks an immediate dollar return through production of abundant food crops destined not for the feeding of the world's starving but for manipulation by politicians seeking to barter for power over people's lives and lands.

Every conscientious homesteader with a far-reaching grasp and a vision of life should entertain the idea of some woodland (tree-covered) acreage within homestead boundaries. The myriad natural plants, animals, birds, and smaller creatures—insects, bacteria, fungi—create numerous harmonies within the woodland expanse. Besides providing aesthetic enjoyment, the balanced, diversified environment of woodlands helps to control soil erosion, provide protection from strong winds, shelter wildlife, and produce firewood, salable timber, and materials for homestead building projects.

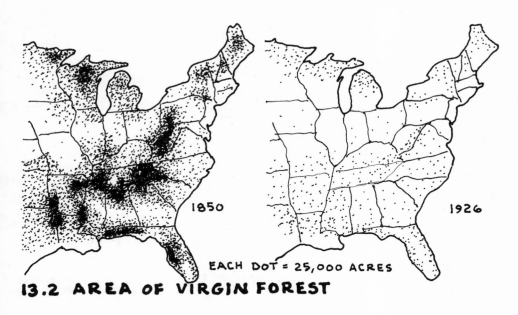

1850

1926

EACH DOT = 25,000 ACRES

13.2 AREA OF VIRGIN FOREST

Woodlands influence climate as well. Winds carry moisture lifted by the sun from large bodies of water until it reaches a wooded area. Trees, creating an obstruction to the movement of moisture-laden air, invite rain to fall in their vicinity. At the same time, transpiring trees cool the air and spray the sky with "upward rainfall." This multiplies the cloud cover and promotes showers.

The soil and moisture conditions required and tolerated by various tree species differ markedly. To make you aware of this, here are some striking contrasts. Yellow poplar requires a deep, moist soil, while black locust will thrive in soil that is either deep or shallow, moist or dry. Alders survive in wet, un-drained soil, while willows require conditions that are wet but well drained. Eucalyptus will dry up a swamp, but it will also drain ground moisture from water-needy areas. Where soil has been damaged through erosion, compaction, or loss of humus, the homesteader should plant conifers. In time they will build up the soil to where hardwoods can be rees-tablished.

The capacity of a species to survive some shade is a factor that will determine the kinds of trees you may choose to propagate in a diversified woodland. The range of shade tolerance is wide, as Figure 13.3 indicates. Seedlings are

TOLERANT	INTERMEDIATE	INTOLERANT
CEDAR	DOUGLAS FIR	PINE
HEMLOCK	ASH	LARCH
REDWOOD	BIRCH	ASPEN
SPRUCE	CHESTNUT	BLACK WALNUT
BEECH	ELM	HICKORY
MAPLE	OAK	WILLOW

13.3 RELATIVE TOLERANCE & GROWTH

RAPID	MODERATE	SLOW
EUROPEAN LARCH	DOUGLAS FIR	CEDAR
LOBLOLLY PINE	PONDEROSA PINE	HEMLOCK
ASPEN	REDWOOD	LONGLEAF PINE
BLACK LOCUST	SPRUCE	BEECH
COTTONWOOD	BLACK WALNUT	OAK - BLACK, WHITE
WILLOW	ELM	SUGAR MAPLE

generally more tolerant of shade than mature trees, especially if you grow them in a congenial environment, with nourishing soil and good exposure to sun. Seedlings grown in proximity to shade givers receive valuable support from the physical protection of the larger trees. Sugar maple seedlings require only 2 percent sunshine for satisfactory growth, whereas loblolly pine seedlings require virtually full sunlight for their development.

Tree root systems have three major shapes: a spear shape, as found in oak, which taps minerals from great depths; a heart shape, as found in the birch tree, which lifts huge quantities of water to tree tops; and a flat shape, which is designed for the support required by such large, heavy trees as the Sitka spruce. It becomes obvious that a tree plantation devoted to a single species, with roots all growing at the same level in the soil, promotes fierce competition for food, moisture, and support.

Homesteaders should encourage a healthy mixture in their woodlands, different varieties of evergreens and deciduous trees, with wide differences in age and development. Mixed woodlands are less subject to insect damage, because they develop a mixed humus. A loose layering of mixed humus creates an environment conducive to only limited support of disease entities that thrive on the products of a particular tree. All over the East, roadside plantings of elm were decimated because of the failure of residents there to realize this fact.

Before delving into the economics of woodland management, we should say something of woodlands as sheltering devices. Our interest in windbreaks was first aroused through reading about the Lake States Forest Experiments at Holdrege, Nebraska. In these experiments two identical test houses precisely recorded their fuel requirements. One house was exposed to winds and one was protected by a windbreak of trees. Both houses were maintained at a constant 70-degree inside temperature. The house with the windbreak required 30 percent less fuel. It was also found in these experiments that during a mild winter animals in a tree-protected yard gained 35 percent more weight.

You must carefully design shelter-belt plantings, keeping in mind tree varieties, planting layout, and tree spacing. For example, poplars are often used as windbreaks because they can withstand pressure from high wind. But poplars are great soil robbers, so it is wise to alternate poplars with a complementary planting of alders. The alder brings nitrogen into the soil, being one of the few leguminous trees that have this important faculty. For reasons mentioned earlier, it is advisable to combine coniferous and deciduous trees. A dense planting of conifers may successfully break the force of prevailing winter winds, but at the same time it will obstruct the flow of cold air, impeding natural air drainage. The winter-bared branches of deciduous trees will not obstruct this necessary movement of air.

The most effective form for a windbreak is an L shape, with the angle pointing in the direction from which the prevailing winter winds originate. This arrangement prevents soil moisture from evaporating, thereby elevating soil temperature; it also discourages snow from drifting across walks and around buildings. Snow in the shade of shelter trees melts slowly in the spring, conserving ground moisture. Row and sod crops benefit by the close proximity of windbreaking woodlands.

Windbreaks, hedgerows, and forests afford safe vantage points from which beneficial birds may survey adjacent croplands. For instance, buzzards, owls, and sparrow hawks cannot check the damaging activities of the sparrow without the protection of a good stand of trees. Attracted to

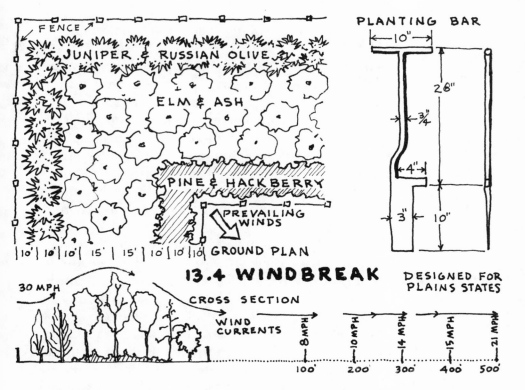

FENCE
JUNIPER & RUSSIAN OLIVE

ELM & ASH

PINE & HACKBERRY

PREVAILING WINDS

|10'| 10'| 10'| 15' | 15' | 10'| 10'|10'| GROUND PLAN

PLANTING BAR

← 10" →

28"

¾"

←4"→

3" ← 10"

13.4 WINDBREAK

DESIGNED FOR PLAINS STATES

CROSS SECTION

30 MPH

WIND CURRENTS

8 MPH   10 MPH   14 MPH   15 MPH   21 MPH

100'   200'   300'   400'   500'

woodlands, birds are important to the control, not the eradication, of destructive congregations of insects. In summer the tomtit has been known to consume 80 pounds of agriculturally injurious caterpillars. The starling consumes harmful larvae. Thrushes, jays, and starlings are all valuable for scattering tree seed.

To encourage bird habitation, you should plant shrubs such as the multiflora rose, especially where fencing is required. You should erect some form of fencing, such as shrubs, woods, or wire, around the perimeter of a woodlot. Grazing a woods may reduce the hazard of fire, but animals, especially sheep and goats, destroy young shoots and compact the humus layer. Installing fence posts and fencing is discussed later.

When contemplating the production of high-grade wood products, keep in mind that you will perform this work during slack time—in winter when homestead chores are at a minimum. Understandably, homesteaders seek for their efforts the greatest quantity of high-grade products available from their wood supply. Such products might be black-walnut furniture, hardwood veneer, saw logs, building poles, or pilings. Low-grade wood products, such as railroad ties, pulpwood, or fuel wood, are culled from the production of high-grade products.

The practice of selling stumpage (standing timber) to a logging contractor is inadvisable in most cases. Homestead-size woodlands only attract the small operator who is notoriously destructive in removing timber. Besides, the sale of stumpage nets the homesteader about 15 percent of the market value of timber. About 30 percent of the operation is charged to felling and bucking the trees. Another 40 percent is taken up in yarding and hauling. The small balance remaining is contractor profit. Even an honest operator will find it impossible, especially when logging a small woodlot, to compensate the landowner for the true value of each log as determined by its grade.

Only when the homesteader harvests his own timber does the high-grading of logs (getting the highest use and value from each log) become possible and feasible. And there is no part of the harvesting that homesteaders themselves cannot do. In due course you will learn that a tall, straight, well-tapered southern yellow pine or Douglas fir will be more valuable as a pole than as a saw log. Large, high-grade logs are best used as veneer wood. Low-grade timber can at first be used as saw logs and ultimately as pulpwood, fence posts, or firewood. Homesteaders can supplement their income by using wood resources as a catch (or cash) crop, just as certain agricultural crops are used to supplement food crops. Unlike many members of society, the homesteader generally does not have access to steady wages, rent, interest, or unearned profit.

The demand for saw logs is greater than for any other timber product, and a well-managed homestead woodlot can produce a minimum of 500 board feet of lumber per acre per year. The key to such production is, of course, in the term *well-managed*. According to the U.S. Department of Agriculture, a woodlot of healthy stock that is managed to its optimum degree will produce three times as much as the average untended woodlot. Three activities are essential to optimum maintenance: (1) choosing the best tree species; (2) keeping the stand at the most favorable density; and (3) cutting and pruning properly.

A conifer is better for wood construction because it is softer wood and easier to work. Softwoods especially valuable for building purposes are pine, spruce, hemlock, and Douglas fir. If you check current nationwide lumber prices, you will find that hemlock and beech have very low stumpage value, whereas birch and maple have a very high value. There is an even greater price range for various grades of lumber. Select white pine sells for $300 per thousand board feet, while number 4 common grade sells for $100.

You can most easily develop a high grade of lumber by maintaining the woods at proper density. The younger the trees, the greater the density. When forests are planted, about 1,000 trees per acre are usually set out. The mature crop contains only about 200 trees. Proper management strives for a uniform rate of growth. If trees are sparsely planted, they grow too fast, increasing the taper and causing larger knots to develop. At proper density the lower branches of a young tree will die and break off, reducing the occurrence of knots. Lumber without knots brings three or four times more money when marketed and is infinitely more valuable and workable for construction purposes. Overcrowding, however, makes a stand grow too slowly, resulting in wood that is light and weak. Where structural strength is required such wood is poor building material.

Pruning, thinning, and cutting, when correctly done, constitute the final factors in optimum woodland maintenance. The aim of thinning is to encourage a tree's fast-growing diameter and its slow-growing height. Consequently, it is better to thin moderately at frequent intervals than to thin heavily infrequently. You should prune and thin in early spring, just before the growing season begins.

When to harvest needs thought. A log 4 feet in diameter contains many board feet of lumber but is not economical to grow. Small trees increase in volume of board feet more rapidly than large ones. As a tree grows from 10 to 11 inches in diameter, the volume of board feet is increased by 33 percent, but with growth from 20 to 21 inches, the volume increases by only 10 percent.

Before attempting to fell a mature tree, consider the damage to surrounding stands that may occur as the tree is on its way down. Keep in mind the slope of the ground, the lean of the tree, wind movement, and the final positioning of the tree for its removal from the woods. Make an undercut first to guide the direction of tree fall. A conifer should be undercut about one-quarter of the diameter of the stump; a hardwood tree should be undercut one-third of its diameter. Make the final cut (or fell) on the opposite side of the tree, slightly above the base of the undercut.

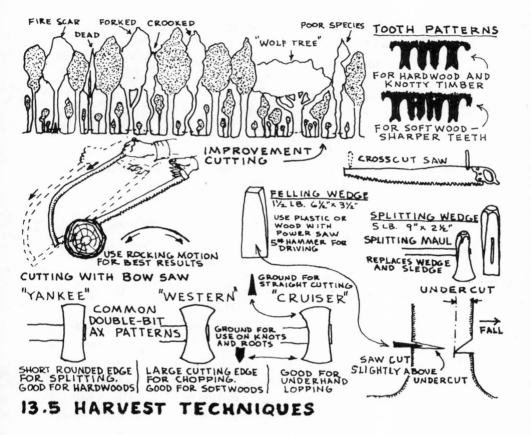

FIRE SCAR FORKED CROOKED

DEAD

"WOLF TREE"

POOR SPECIES

TOOTH PATTERNS

FOR HARDWOOD AND KNOTTY TIMBER

FOR SOFTWOOD — SHARPER TEETH

IMPROVEMENT CUTTING

CROSSCUT SAW

USE ROCKING MOTION FOR BEST RESULTS

CUTTING WITH BOW SAW

FELLING WEDGE
1½ LB. 6½" x 3½"

USE PLASTIC OR WOOD WITH POWER SAW 5# HAMMER FOR DRIVING

SPLITTING WEDGE
5 LB. 9" x 2½"

SPLITTING MAUL

REPLACES WEDGE AND SLEDGE

"YANKEE"

"WESTERN"

GROUND FOR STRAIGHT CUTTING

"CRUISER"

UNDERCUT

COMMON DOUBLE-BIT AX PATTERNS

GROUND FOR USE ON KNOTS AND ROOTS

FALL

SHORT ROUNDED EDGE FOR SPLITTING. GOOD FOR HARDWOODS

LARGE CUTTING EDGE FOR CHOPPING. GOOD FOR SOFTWOODS

GOOD FOR UNDERHAND LOPPING

SAW CUT SLIGHTLY ABOVE UNDERCUT

13.5 HARVEST TECHNIQUES

On the assumption that a picture is truly worth a thousand words, Figure 13.5 should present the final statement necessary to acquaint the homesteader with the important facets of woodland harvesting.

Fuel wood, one of the by-products of a well-managed woodlot, is a suitable cottage industry for the homestead, bringing in extra cash income. From thinnings and prunings alone you can realize about 1 cord of new wood per acre per year. A cord of hardwood weighs 2 tons, twice the weight of softwood, and, when seasoned (dry), will give as much heat as 200 gallons of fuel oil or 1 ton of the very best anthracite coal. A stack of cordwood measures 4 feet x 4 feet x 8 feet, or 128 cubic feet, and will yield a hundred 4-hour fires in an efficient wood-burning stove.

Maple sugar production is a practical cottage industry for those living throughout the northeastern United States. We urge anyone interested in tapping sugar maples to read Scott and Helen Nearing's *Maple Sugar Book*. It contains everything you need to know on the subject, as well as some beautiful and well-documented historical commentary. One of their references, Franklin Howe, succinctly summed up this sugary subject in 1884:

> The maple grove that is planted by a young man may be enjoyed by him through more than half of an ordinary lifetime. With proper care it will perpetuate itself through a long course of years, and for aught we know (if the young growth is protected) forever. It will occupy broken grounds that could not otherwise be cultivated, and the timber, when taken out at greatest maturity, has a value which is gaining every year, aside from the annual revenue to be derived from the sap. The maple adorns and beautifies perhaps more than any other of our native forest trees. . . . The sugar season comes at a time when farm labor is least employed, and the occupation presents amenities beyond those which any other form of farm labor can afford.

In closing we offer fair warning about the current rash of get-rich-quick schemes involving woodland production. The 1970s craze for Christmas-tree cultivation reminds us of the chicken-raising and mushroom-growing enterprises of the 1940s, the worm-cultivating and rabbit-breeding projects of the 1950s, and the chinchilla-producing and herb-farming rackets of the 1960s. To this day we can readily find concrete blocks for fireplace construction in the rubble of long-abandoned pen construction for abortive nutria-raising schemes of the 1940s.

Many entrepreneur homesteaders have lost their entire life savings in these projects. There is obviously cause for hesitation when considering Christmas-tree production as a cash crop. It has been calculated, for instance, that on an acre of submarginal land a person can raise a hundred Christmas trees per year. On this basis, in any year 600,000 acres will supply all of the trees that might conceivably be desired by the entire United States. But with 50 million acres of the country's submarginal land actually suitable to this kind of enterprise, the homestead entrepreneur of Christmas-tree cultivation will likely find the market for these trees diluted and the profits in general low. The homesteader who invests in Christmas-tree-growing can expect for his considerable efforts and investments a profit of $1 from each balsam fir at the end of a ten-year growing period. But a loblolly pine that is cut for piling will net a $50 profit after sixty years' growth. Homesteaders stand to lose more than money from Christmas-tree production. From our point of view they denigrate a potentially beautiful tree-growing experience for the profit from a pagan ritual.

The immediate economic advantage of salable timber may appeal to many, but eventually you will gain greater return through proper management and the production of quality finished wood products. The homestead, meanwhile, enjoys the benefits of the living trees: windbreaks, conservation of soil and water, refuge for wildlife, and the immeasurable enhancement of everyone's aesthetic sensibilities.

A profile sketch of the authors' family homestead showing the location of native tree types, elevation, and rainfall appears in Figure 13.6. When we view our hills and woods, our homestead site gains meaning and significance.

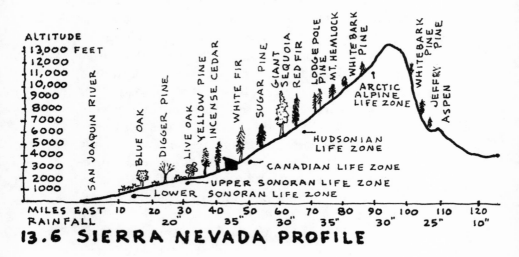

13.6 SIERRA NEVADA PROFILE

We should mention that, of all the governmental bureaucracies, state forestry officials remain the most helpful and cooperative. You can obtain sound advice or sources for assisting further development of individual forestry projects by writing your state forester:

Alabama: State Forester, 64 N. Union St., Montgomery 36104
Alaska: State Forester, 344 Sixth Ave., Anchorage 99501
Arizona: State Forester, 422 Office Bldg. E, Phoenix 85005
Arkansas: State Forester, P.O. Box 1940, Little Rock 72203
California: State Forester, Resources Bldg., Sacramento 95814

Colorado: State Forester, Colorado State University, Fort Collins 80521

Connecticut: State Forester, 165 Capitol Ave., Hartford 06103

Delaware: State Forester, 317 S. State St., Dover 19901

Florida: State Forester, P.O. Box 1200, Tallahassee 32304

Georgia: Forestry Director, P.O. Box 1077, Macon 31203

Hawaii: State Forester, 400 S. Beretania St., Honolulu 96813

Idaho: State Forester, 801 Capital Blvd., Boise 83701

Illinois: State Forester, 400 S. Spring St., Springfield 62706

Indiana: State Forester, 607 State Office Bldg., Indianapolis 46209

Iowa: State Forester, E. 7th & Court Sts., Des Moines 50309

Kansas: Ext. Forester, Kansas State College, Manhattan 66504

Kentucky: Forestry Director, New Capitol Annex, Frankfort 40601

Louisiana: State Forester, P.O. Box 15239, Baton Rouge 70815

Maine: Forest Comm., Maine Forestry Dept., Augusta 04330

Maryland: State Forester, State Office Bldg., Annapolis 21401

Massachusetts: Forestry Director, 15 Ashburton Pl., Boston 02108

Michigan: State Forester, Steven T. Mason Bldg., Lansing 48926

Minnesota: Forestry Director, Centennial Office Bldg., St. Paul 55101

Mississippi: State Forester, 1106 Woolfolk Bldg., Jackson 39201

Missouri: State Forester, P.O. Box 180, Jefferson City 65102

Montana: State Forester, 2705 Spurgin Rd., Missoula 59801

Nebraska: State Forester, University of Nebraska, Lincoln 68508

Nevada: State Forester, 201 S. Fall St., Carson City 89701

New Hampshire: Resident Director, State Office Bldg., Concord 03033

New Jersey: State Forrester, Labor and Industry Bldg., Trenton 08611

New Mexico: State Forester, P.O. Box 2167, Santa Fe 87501

New York: Director, Division of Lands and Forests, Albany 12183

North Carolina: State Forester, P.O. Box 2719, Raleigh 27602

North Dakota: State Forester, State School of Forestry, Bottineau 58318

Ohio: Chief, Forestry Division, 815 Ohio Depts. Bldg., Columbus 43215

Oklahoma: Forestry Director, Capitol Bldg., Oklahoma City 74074

Oregon: State Forester, P.O. Box 2289, Salem 97310

Pennsylvania: State Forester, Bureau of Forests, Harrisburg 17120
Puerto Rico: Commonwealth Forester, P.O. Box 10163, Santurce, San Juan
Rhode Island: Chief Forester, 83 Park St., Providence 02903
South Carolina: State Forester, 5500 Broad River Rd., Columbia 29202
South Dakota: State Forester, Pierre 57501
Tennessee: State Forester, 2611 W. End Ave., Nashville 37203
Texas: Director, Texas Forest Service, College Station 77843
Utah: State Forester, 525 W. 1300 South, Salt Lake City 84115
Vermont: Director of Forests, Montpelier 05601
Virginia: State Forester, P.O. Box 3347, Charlottesville 22902
Washington: Forestry Supervisor, P.O. Box 110, Olympia 98501
West Virginia: State Forester, Department of Natural Resources, Charleston 25305
Wisconsin: State Forester, P.O. Box 450, Madison 53701
Wyoming: State Forester, Capitol Bldg., Cheyenne 82001

14
The Pit Greenhouse

When you understand how to do a thing, the doing is easy; if you find it difficult you do not understand it.
                                                            Persian proverb

There are a number of homestead activities about which a basic understanding can make the difference between a project's being simple or difficult, a gratifying success or a disheartening failure. Nowhere on the homestead is this dichotomy more evident than when you attempt to modify the plant-growing environment by building and operating a greenhouse.

Some types of plant shelter are built to control only one or another factor of that environment, such as excessive heat, wind, or cold. To temper these extremes, gardeners often build a shading device, a windbreak screen, or even a rudimentary cold frame. These are all rational structures requiring minimum knowledge to construct and operate. But attempting to modify the entire growing climate in a greenhouse is a more complex undertaking.

Both George Washington and Thomas Jefferson built contemporary horticultural circles, yet it is an imaginative dard structure was a pit greenhouse built into a Waltham, Massachusetts, hillside about 1800. Covered on its south side with glass and sometimes quaintly called a glasshouse, it was sun-heated and vented. Its cold north side was insulated by an earth berm. The traditional greenhouse is built entirely above ground.

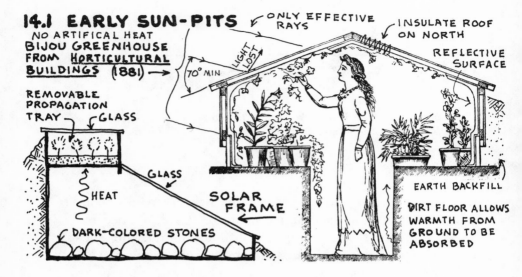

14.1 EARLY SUN-PITS

ONLY EFFECTIVE RAYS

NO ARTIFICAL HEAT
BIJOU GREENHOUSE FROM HORTICULTURAL BUILDINGS (1881) →

70° MIN

LIGHT LOST

INSULATE ROOF ON NORTH

REFLECTIVE SURFACE

REMOVABLE PROPAGATION TRAY
GLASS

GLASS

HEAT

SOLAR FRAME

DARK-COLORED STONES

EARTH BACKFILL

DIRT FLOOR ALLOWS WARMTH FROM GROUND TO BE ABSORBED

This design seems to be either unknown or ignored in contemporary horticultural circles, yet it is an imaginative and efficient solution to the need for a temperature-modifying, plant-forcing structure on the homestead. As presently built, this structure is large enough to permit the gardener to stand at work-height benches, 1 foot deep and 3 feet wide, which contain fertile, friable soil. Its southern exposure is covered with a translucent material that admits the radiant energy of sunlight. This is vital to plant development, for in the absence of this light energy, plants manufacture no food and starve.

The greenhouse is an enclosure that modifies the changes in temperature between night and day, summer and winter— a temperature equalizer, if you will. Heat rays from the sun penetrate during the day, warming its interior. This heat is trapped within, escaping slowly and moderating the cold of night, the chill of winter. Temperatures above or below certain limits (50 to 90 degrees Fahrenheit for warm-weather vegetables and 40 to 80 for cool-weather vegetables) stall plant development. If you grow plants under glass for early transplantation to outside plant beds, you should provide them with temperatures near those indicated in Figure 14.2. Too high a temperature will produce spindly plants. Watch the type of growth closely, and adjust the temperature accordingly.

Humidity and carbon dioxide are other factors of the greenhouse environment that must be controlled. Handling all these factors properly creates greenhouse conditions that promote the growth of seedling transplants early in the growing season, many weeks before the outside environment is entirely hospitable to them. In the United States the season for outside gardening can range from as little as three to as many as eight months between February and October.

Sunlight is the energizer for all plant growth, inside a greenhouse as well as outside in the garden. It is made up of energy bursts of varying length that radiate from the sun, filling the universe. This electromagnetic radiation originates as vibrations emitted from the atoms and molecules of exploding, burning gases at the sun's surface. Reaching earth's atmosphere, some of these emissions pass through it; the rest are filtered out by the ozone layer. On either side of the visible wavelengths of white sunlight are invisible wavelengths. At one extreme is a wavelength called ultraviolet. Fortunately, many of these emissions do not pass through our airy en-

14.2 TEMPERATURE RANGE

	CROP	DAY	NIGHT
COOL	BROCCOLI	60-65	55-60
	BRUSSELS SPROUTS	60-65	55-60
	CABBAGE	55-60	50-55
	CELERY	65-70	60-65
	LETTUCE	55-60	50-55
	ONION	60-65	55-60
WARM	EGGPLANT	70-80	65-70
	MUSKMELON	70-75	60-65
	PEPPER	65-70	60-65
	TOMATO	65-70	60-65
	WATERMELON	70-80	65-70

velope, for continual exposure to them would harm us. A portion of this wavelength, however, does reach earth as a part of natural outdoor daylight and is beneficial to certain life processes. The number of these longer ultraviolet wavelengths is somewhat less than 7 percent of the total useful sunlight we receive, yet according to light scientist Dr. John Ott, these select rays have been instrumental to life's evolution on earth. Essential to chloroplast movement in leaf cells, long-wave ultraviolet light is also necessary to the production of vitamin D, so vital to bone growth in animals.

Another element of sunlight that is also invisible to us but of which we are sometimes uncomfortably aware is the infrared or heat waves. These are the wavelengths that cause daytime heat buildup in our environment. Striking earth's surfaces, they are reradiated into earth's atmosphere, from which they slowly escape back into outer space. When the northern hemisphere is more nearly perpendicular to these incoming rays, as it is in summertime, it receives more of their intensity, and plant metabolic processes are speeded up, "bringing out the sugar" in ripening fruit. As far as the greenhouse is concerned, this heat must be moderated while we seek light in abundance, full natural sunlight, the catalyst for the magic of plant chemistry and the evolution of life.

Photosynthesis literally means "putting together in light." By the catalytic stimulus of sunlight, carbon dioxide and water in the presence of the green chlorophyll in leaves are converted to the simple sugar from which plants manufacture their food and structural tissue, cellulose. When animals consume these materials, plant tissue is converted to animal tissue, eventually returning to the environment again as carbon dioxide and water. Through this progression a balance is maintained between plant and animal life, and the energy of the sun is miraculously made available for the energy needs of the animal organism.

Besides adequate moisture and minerals in sufficient quantities from fertile, friable soil, carbon dioxide is the most important ingredient taken in by plants. Trees and plants must sift great volumes of atmosphere for this airy fertilizer, which they take in during daylight hours and release at night, when photosynthesis stops. Depending on their rate of photosynthesis, greenhouse plants need and use much carbon dioxide (more on bright, sunny days than on overcast days).

Greenhouses need additional sources of carbon dioxide. In warm climates, high concentrations cannot be maintained as long as the greenhouse is controlled by ventilation; air movement disperses carbon dioxide to the outside atmosphere. In the closed winter greenhouse, carbon dioxide levels can be increased without this dissipation, although there is some net reduction due to ongoing photosynthetic activity. Plants return some carbon dioxide to the air with their respiration, but more is still necessary to supply spectacular results. One enterprising couple we know raises rabbits in cages beneath their greenhouse benches. They report increases of up to 800 parts per million of carbon dioxide, as well as a harvest of nitrogen-rich manure and delectable meat. Both animals and plants appear to thrive in this complementary environment.

Many greenhouse coverings allow most of the infrared heat waves to penetrate the greenhouse, where their wavelength is altered as it is reradiated from plants, soil, and interior surfaces. These lengthened wavelengths cannot as readily pass back through some covering materials. Their heat is therefore trapped within the enclosure, intensifying inside temperatures. High levels of both carbon dioxide and water vapor also obstruct the passage of reradiated heat waves through the greenhouse envelope. This entrapment is called the greenhouse effect. Desirable in winter to modify inside-outside temperature differences, this effect must be carefully monitored and moderated during warmer seasons, when respiration and metabolic processes will be unnaturally speeded up, causing plants to "burn."

Although the gardener seldom uses a greenhouse during the hot summer season there are days in winter when temperatures climb within this enclosure. A pit greenhouse built by Helen and Scott Nearing for their mountainside homestead in Vermont was in active use most of the year for the several decades of their residence there. They built the greenhouse for raising transplants for a summer garden of short duration, but they soon found that this plant-forcing structure functioned well as an all-winter source of various fresh greens, even with no supplemental heat for the often subzero temperatures. "On mild, sunny days in winter," they write, "with no stove or artificial heating, the temperature inside this sun-heated greenhouse went up to 100 degrees unless we ventilated it."

Over the winter, the Nearings also found that if they transplanted fall lettuce to the unheated pit, it survived temperatures down to 25 degrees below zero, even though it froze stiff. If left to thaw on a warmer day, the lettuce soon "stood in the greenhouse crisp and edible." They felt that had their greenhouse been roomier, they might have grown mustard greens, garden cress, leaf chicory, and turnip greens all winter with equal success.

Controlling season-long subzero temperatures is another matter. During the coldest months of winter the intensity and duration of sunlight are at their lowest levels. This requires a design that will reflect as much light as possible down onto growing plants, a design such as that sketched in Figure 14.3 by itinerant greenhouse designer Bruce Bugbee.

In regions affected by severe winter cold, some manner of insulating cover is required at night, one that thoroughly

14.3 "MORNING GLORY" GREENHOUSE

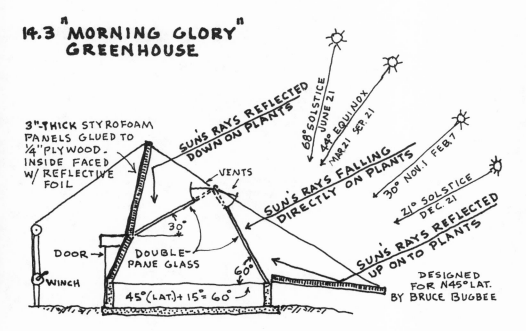

3"-THICK STYROFOAM PANELS GLUED TO ¼" PLYWOOD. INSIDE FACED W/ REFLECTIVE FOIL

SUN'S RAYS REFLECTED DOWN ON PLANTS

VENTS

68° SOLSTICE JUNE 21

44° EQUINOX MAR 21 SEP. 21

SUN'S RAYS FALLING DIRECTLY ON PLANTS

30° NOV.1 FEB.7

21° SOLSTICE DEC. 21

SUN'S RAYS REFLECTED UP ON TO PLANTS

DOOR

DOUBLE-PANE GLASS

30°

60°

WINCH

45° (LAT.) + 15° = 60°

DESIGNED FOR N45° LAT. BY BRUCE BUGBEE

seals off the greenhouse from any chilling air leaks. A covering of a double layer of transparent material may be sufficient for less severe climates. If carefully sealed, this layering creates a dead-air space, usually 1 centimeter thick. Polyethylene film may suffice when stretched over the inside or the outside of the greenhouse cover and may well cut heating requirements in half. Auxiliary heat, such as that from Riteway's thermostatically controlled wood stove connected to a passive system of heat storage, may be necessary.

In northern climates when the sun is both lower in the sky and farthest away from us, the greenhouse wall facing south should be at an optimum angle perpendicular to the sun. According to the formula, this would be your degree of latitude plus 15. Fifty-five degrees is suggested for the slope of the north wall, which receives reduced light. The sun's altitude at

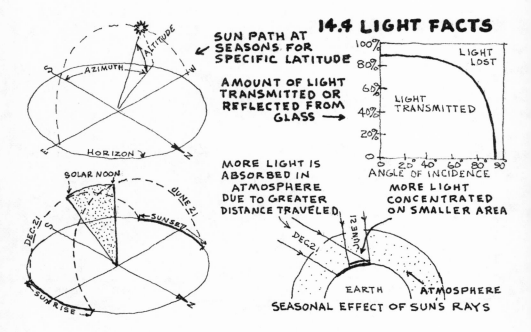

14.4 LIGHT FACTS

SUN PATH AT SEASONS FOR SPECIFIC LATITUDE

AZIMUTH

ALTITUDE

HORIZON

SOLAR NOON

DEC. 21

JUNE 21

SUNSET

SUNRISE

AMOUNT OF LIGHT TRANSMITTED OR REFLECTED FROM GLASS →

LIGHT LOST

LIGHT TRANSMITTED

ANGLE OF INCIDENCE

MORE LIGHT IS ABSORBED IN ATMOSPHERE DUE TO GREATER DISTANCE TRAVELED

MORE LIGHT CONCENTRATED ON SMALLER AREA

DEC. 21

JUNE 21

EARTH

ATMOSPHERE

SEASONAL EFFECT OF SUN'S RAYS

this time is subject to geographical variation; it is approximately plus or minus 1 degree for every degree of north or south latitude. Therefore, in Seattle, Duluth, or Bangor the winter sun will reach an altitude of 19 degrees, while the summer sun will climb to an altitude of 66 degrees.

Another factor—greenhouse condensation—requires ventilation control. As the temperature in the enclosure rises, warm water vapor condenses, forming liquid droplets on the translucent covering in contact with the cold outer air. This not only reduces greenhouse humidity but diminishes light transmission by half, and the dripping often spreads plant disease. Therefore, you must provide adequate ventilation to control these variables; allow for plenty of air movement, as you can always close it off later if it is too much. A rule of thumb is that the ridge vent should be one-eighth the width of the greenhouse. An 8-foot-wide structure would require a ridge vent 1 foot wide. For smaller greenhouses open doors and end vents may suffice, allowing smaller ridge vents.

Several light sprinklings early and late in the day are usually all that are required to maintain optimum humidity. It is far better to sprinkle greenhouse floors and only the soil in the benches, for water on plants opens their stomata, causing them to transpire more rapidly. Increased transpiration is the result of high daytime temperatures, causing great plant stress. It can be reduced if daytime humidity is kept correspondingly high. Optimum daytime humidity for tomato plants, for example, is from 50 to 80 percent, while the nighttime humidity should be 95 percent. Daytime temperature requirements of these plants are 65 to 70 degrees, with nighttime temperatures of 60 to 65 degrees.

It is not necessary to control the humidity and temperature of the greenhouse with ventilation operated by thermostats, clocks, and switches. Again, greenhouse operation is an intimate affair between life forces and their human attendants. Heat and moisture can easily be monitored by personal, daily observation of growing plants and by several strategically placed thermometers. Vents and open doors circulate air. Sometimes in hot weather you may need to use shading devices—bamboo slats, airy quilts of burlap stuffed with leaves, or lightweight tarpaulins. In very warm climates you may plant deciduous trees near the structure on its hot, western side.

Designers are never quite in agreement about the proper direction in which to orient the greenhouse. Some operators feel that morning hours are the most productive for the manufacture of plant food, that in the afternoon the energy of the plant has started to wane, interrupting further food production. In this case a southeastern exposure would be the best. On the other hand there are those who feel that light is light, that plants do not care if it is morning or evening light. For northern climates a southeasterly exposure would allow the sun to thaw ice and condensation from the greenhouse covering earlier on a winter morning, admitting more light sooner. Insulated panels would, of course, keep this ice from forming in the first place. Plant energy would only wane in the afternoon if temperatures were allowed to build too high, causing plants to wilt.

Glass often covers the standard commercial greenhouse today. It is more durable than any other covering. While admitting many wavelengths, ordinary crystal window glass refracts (scatters) or filters out most beneficial ultraviolet radiation. This creates a health hazard for man as well as plants. Painting glass with whitewash to repel infrared heat waves also repels this beneficial ultraviolet radiation. The obvious solution is to use a type of translucent material for the greenhouse covering that will admit all wavelengths of the maximum available sunlight.

The most common material used today for greenhouse covering is polyethylene film, which transmits most of the useful solar energy. Infrared heat waves readily enter and leave through this membrane, causing rapid daytime warming and rapid night cooling. It is therefore not uncommon for a polyethylene-covered greenhouse to have a lower inside nighttime temperature than that found outside. To prevent low temperatures from destroying your greenhouse crop, you must protect this covering with additional insulation and possibly an auxiliary heating system.

Unfortunately, ultraviolet light causes polyethylene film to disintegrate after it has been exposed to sunlight for a year. The plastics industry has therefore added an ultraviolet inhibitor to this film so that it will last a year longer. Another, heavier film, polyvinyl chloride (PVC), may last twice as long as regular polyethylene film but will admit only a fraction of the amount of infrared radiation. The amount of heat admitted during the day and lost at night is less than when polyethylene is used. Other films that you can use are the polyester Mylar, which lasts up to four years, and the polyvinyl fluoride Tedlar, which may be serviceable up to eight years. These plastics are also available in the form of panels.

Anyone seriously involved in greenhouse plant production will not be satisfied with short-lived film coverings. The ideal covering should transmit a high degree of sunlight and be rigid, strong, and resistant to weathering. Fiber-glass-reinforced plastic panels meet these specifications. Some types transmit as much as 86 percent of the solar light coming to earth, comparable to glass, which transmits 90 percent. Fiber glass will not soften when heated, and although it weighs only 5 ounces per square foot, it will support 100 pounds per square foot. Some types will last approximately twenty years, at which time you may recoat them by brushing on resin.

Polyesters have high reflectivity; their large number of glass fibers break up and diffuse incoming light. The result is a uniform light level. This also makes fiber glass initially slow to heat up at the beginning of a sunny day, reducing its potential greenhouse effect. Heat loss is one-quarter that of glass and thirty-five times less than that of the plastic films.

Major heat and light loss occurs through the larger surface area of corrugated plastic panels. Flat sheets conserve one-third more heat and light. When dust and soot settle on corrugations, static charges build up on the covering. Other contaminants are then attracted to this surface, drastically reducing light transmission.

Light, as we have seen, is the energizer for the plant-growing process known as photosynthesis. The human eye sees visible sunlight as white. However, when passing through a transparent refracting device, like a glass prism or a drop of water, sunlight demonstrates a range of color made up of the various hues of the rainbow: violet, blue, green, yellow, orange, and red. Recent experiments in Japan show that blue light stimulates plant growth, while green light inhibits it. The Japanese even use green mulching films to prevent weed growth. They also manufacture a PVC film capable of passing selected wavelengths of light, while excluding others. Blue light, they found, increases the number of flowers on strawberry plants and the vitamin C content of parsley.

At the other end of the spectrum, infrared radiation does little more than heat the environment, as stated previously in this chapter. Chlorophyll absorbs quantities of red light waves, resulting in strong photosynthetic activity. Weak activity occurs in the green part of the spectrum, and strong chlorophyll absorption and photosynthetic activity begin again in the blue range. Beyond the slender range of beneficial ultraviolet radiation, bleaching and leaf burning occur (see Figure 14.4).

Japanese experiments also corroborate that crops grown under polyethylene films or fiber glass mature earlier and are of higher quality and greater yield than those raised under glass. Apparently, as Dr. John Ott, director of the Environmental Health and Light Research Institute (Sarasota, FL), has shown for some years, those beneficial ultraviolet wavelengths near the blue range affect the color, flavor, and texture of fruits and vegetables grown in a greenhouse. Plastic coverings tinted blue will affect the movement of chloroplasts in the cells of plant leaves.

Such were the suspicions that caused General A. J.

Pleasanton to postulate his blue-glass theories in the nineteenth century. This inventive genius built a greenhouse in which every eighth row was covered with glass of a blue tint. At the end of the first five months of plant growth in this environment, grapevines in his 84-foot-long greenhouse produced 1,200 pounds of fruit on vines that reached 45 feet in length. Acclaiming this achievement, Pleasanton wrote:

> That blue light of the firmament, if not itself electromagnetism, evolves those forces which compose it on our atmosphere, and applying them at the season, viz., the early spring, when the sky is bluest, stimulates, after the torpor of winter, the active energies of the vegetable kingdom, by the decomposition of its carbonic acid gas— supplying carbon for the plants and oxygen to mature it and to complete its mission.   (*The Influence of the Blue Ray of the Sunlight and the Blue Colour of the Sky*, 1876)

Indeed, we do know that the blue part of sunlight is usually scattered more than the other colors, and thus, because of atmospheric diffusion, we have a blue sky, under which all life has evolved.

Figure 14.5 illustrates the range of chlorophyll synthesis. Carbon dioxide is absorbed into the plant through the stomata, intercellular openings located in the outer skin of leaves. Oxygen is transpired through these orifices. The stomata open more in the presence of blue and red light than in green light. Chlorophyll formation is accelerated when growing plants are exposed to these colors. The degree of carbon dioxide absorption and chlorophyll formation is significant under blue and red light.

Photosynthetic lighting using fluorescent lamps has been developed to stimulate chlorophyll synthesis. These lamps are

14.5 ELECTROMAGNETIC SPECTRUM

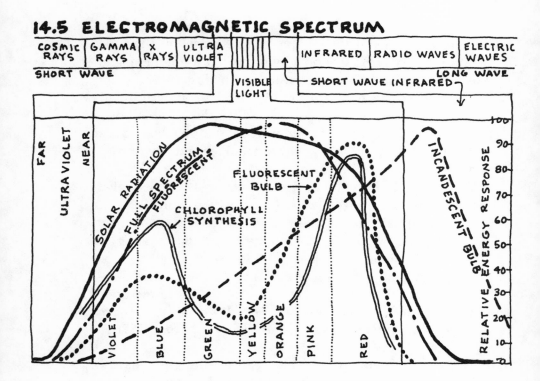

designed to emit energy high in the red as well as the blue of the spectral band. The green and yellow portions of the spectrum are filtered out. Fluorescent lamps are better than incandescent lamps for this purpose. The phosphor coating within the lamp converts short-wavelength radiation into a long-wavelength form, producing four times the light output for each watt input—a ratio considerably higher than that of incandescent bulbs and requiring less power to produce the same amount of light. Incandescent bulbs produce a high output of infrared heat waves, while fluorescent lamps limit it and remain relatively cool. Light from fluorescent lamps is more evenly distributed, and the lamp itself lasts considerably longer than the incandescent.

Sylvania's Gro-Lux fluorescent is probably the best-known wide-spectrum lamp. Vita Lite is another, designed not only for plants but for human vision as well. This lamp is manufactured by Duro-Lite Company, Fair Lawn, NJ. In the homestead residence you would do well to replace all the ultraviolet-inhibiting window glass and incandescent bulbs with transparent plexiglass and full-spectrum fluorescent lamps. At this writing we have information that the Solar Lighting Corporation (1520 West Fulton St., Chicago, IL 60607) has in production a new full-spectrum, radiation-shielded fluorescent fixture. This lamp will provide all the visible wavelengths, plus the proportionate amount of long-wavelength ultraviolet light contained in natural outdoor daylight. Contact the Rohm and Haas Company for Ultraviolet-Transmitting Plexiglass (UVT Plexiglass) or a dealer for American Cyanamid Company and ask for UVT Acrylite, requesting the ultraviolet-transmitting type.

Greenhouse designers seem to agree that a dome is the best shape for catching the greatest amount of natural light during the months of least sunshine. A dome greenhouse is seldom seen, however, for it is unsuitable for commercial purposes, and glass and plastic coverings usually can be used only for square-shaped buildings. Figure 14.6 shows an experimental dome greenhouse we are building on our homestead. This half-dome is aligned along its east-west axis. Its curving vertical outside wall is cast with the same double-wall slip form we will use to build most other structures on our new homestead. Its curving back (north) wall is smooth and painted to reflect any sunlight striking it onto growing plants. Full-spectrum lighting under the growing bench doubles our growing area. Note that a solar-heated draft and wind turbine supply the ventilation.

The fiber glass panels of this greenhouse transfer max-

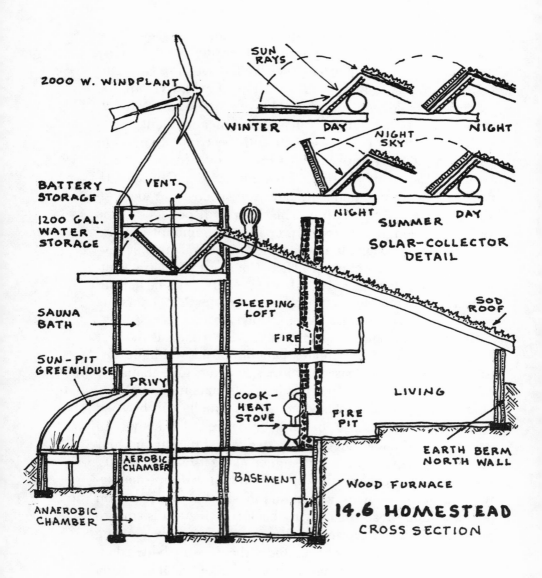

2000 W. WINDPLANT

SUN RAYS

WINTER   DAY   NIGHT

NIGHT SKY

NIGHT   SUMMER   DAY

SOLAR-COLLECTOR DETAIL

BATTERY STORAGE

VENT

1200 GAL. WATER STORAGE

SLEEPING LOFT

FIRE

SOD ROOF

SAUNA BATH

SUN-PIT GREENHOUSE

PRIVY

COOK-HEAT STOVE

LIVING

FIRE PIT

AEROBIC CHAMBER

BASEMENT

EARTH BERM NORTH WALL

ANAEROBIC CHAMBER

WOOD FURNACE

14.6 HOMESTEAD
CROSS SECTION

imum light since the sun's rays are always perpendicular to some portion of this covering, summer or winter. Anchored at their top and bottom and attached at their seams but otherwise self-supporting, these panels are individually cast to fit the vertical arc and the greenhouse diameter. We added a blue cobalt catalyst to the resin to improve spectral and weathering qualities. These unique panels are now manufactured by a California plastics company, and you can obtain further information about them by writing us.

Late in the 1950s *Organic Gardening and Farming* magazine featured an article about an unheated pit greenhouse that we integrated into the interior living space of a California home. With this arrangement a person could stand in the kitchen while picking salad from the greenhouse bench. Growing plants added zest to the whole visual environment—yet another instance of the desirability of close proximity of people and plants. Despite these and earlier references to the intimacy of this relationship, it may well be that this delightful practice significantly increases the level of carbon dioxide, accelerating the rate of plant growth. This may explain why it is that plants seemingly respond so well to talk and prayer; it may not be the sentiment expressed for them that "turns them on," but rather the carbon dioxide we breathe upon them.

15
Fences

Of all occupations from which gain is secured, there is none better than agriculture, nothing more productive, nothing sweeter, nothing more worthy of a free man.   *Cicero*

For thousands of years farmers have erected fences to enclose crops and domesticated animals and to exclude predators. The resultant ordering and securing of the environs makes fence-building one of the sweeter, more productive, and enduring of farmstead activites. Yet this is the kind of endeavor for which homesteaders seldom allocate sufficient time or funds. Consequently, temporary slipshod fencing is a familiar sight. The construction of durable fencing is not unlike the building of other homestead structures. Preparing a thorough plan, selecting appropriate tools and materials, and employing proper procedures for construction is mandatory.

Because of labor investment and the cost of materials (which averages 40 cents per lineal foot), a realistic fence-building program requires that you recognize the priority stages before you begin. Depending on the purpose you intend, the fencing of unimproved homestead land may have as many as seven of these stages. The types of fencing you may need include the following:

1. Boundary fencing. A permanent fence may be desirable to delineate the farmstead perimeter, thereby controlling casual trespass through cultivated areas.

2. Tree-crop fencing. No successful management of tree crops can take place without protection from encroachment by domestic or wild animals.
3. Holding-yard fencing. Some form of enclosure is necessary for the occasional close confinement of farm animals.
4. Garden fencing. Security from animal ingress is requisite to food-crop production.
5. Pond fencing. For an undisturbed pond environment, fencing from livestock is neccessary.
6. Woodlot fencing. A fence around cultivated woods indicates the serious purpose of this enclosure to fire-making picnickers and prevents livestock damage, particularly to young trees.
7. Permanent pasture fencing. The fencing of a number of pasture areas facilitates crop rotation and prevents overgrazing.

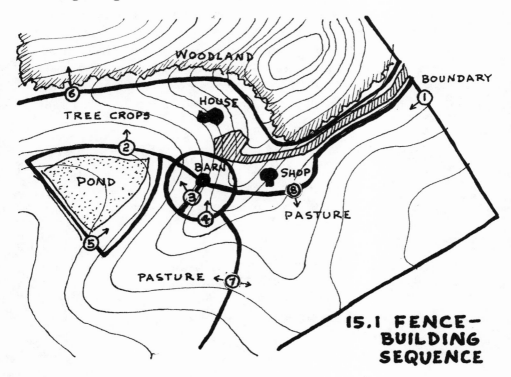

15.1 FENCE-
BUILDING
SEQUENCE

8. Field-division fencing. Grazing areas of similar soil, moisture content, or steepness should be separately fenced. Fencing allows early spring grazing of high land, while it reserves bottomland for summer and fall grazing.

Several benefits may result from building a boundary fence. True property lines and corners will be established when the fence line is surveyed, and as the land is cleared for actual construction, a firebreak and boundary road will result. It is advisable to remove nearby trees whose falling branches or windblown limbs may menace an otherwise well-built fence.

Before you can set the posts and string the wire, you must give some consideration to clearing the land. An investment in a small blade-equipped crawler tractor would solve this problem with dispatch, but there are often inaccessible places on the homestead where you cannot use tractors. It may, therefore, be necessary to grub some fence lines by hand.

A wide variety of hand and power tools is available for clearing land, each designed for a specific purpose. The basic land-clearing tool is the machete, an excellent choice for cutting vines and small brush. You can efficiently remove larger

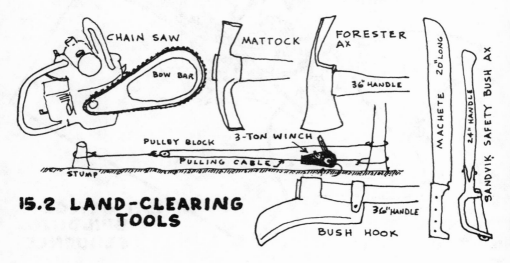

15.2 LAND-CLEARING TOOLS

vines and saplings by using a brush hook. Although handled like an ax, the brush hook is preferable to one. It has a wider blade that wears at the edges, showing little use in its center. The Swedish (Sandvik) safety bush ax has proven faster, safer, and easier to use than the common ax. Thin, flat, replaceable blades cut quickly through young, springy trunks. Few tools can outperform the mattock for general grubbing work. A Pulaski, or forester's, ax, patterned after the mattock, is a practical choice for general clearing work where maneuverability and tool weight are appreciable factors. Finally, to remove larger brush and trees, a homesteader would do well to invest in a gasoline-powered chain saw equipped with a bow bar. The bow bar is specially designed to cut and trim small trees. The narrow rail that closely follows the saw teeth is unlikely to bind in this activity.

The mowing machine is an indispensable tool for fence-line clearing. A well-built mower will cut young brush if it has not gone through its first winter's hardening. The late summer months of August and September are therefore the best time to clear newly sprouted brush in order to retard the sturdier sprouting of the following year.

For the few years that reviving growth persists, browsing stock—such as goats and sheep—will do much by nibbling to eliminate brush and tree shoots. The soil around the roots of such vegetation will be packed by their hooves, causing its decay. If a tree is girdled or treated with Ammate crystals (ammonium sulfate, Heddy Corporation, Paterson, NJ 07524) and left to stand a year before it is cut down, sprouting will not occur later. If you do not keep land free of brush after its initial clearing, in a period of three years there will be more brush than you previously encountered.

Some success has been achieved by sowing newly cleared land with a mix of ryegrass and crimson clover. No special soil preparation, such as tilling, is necessary. Kudzu grass is another sod crop used for clearing land. Its luxuriant growth will rot stumps and actually choke out brush and trees.

To fence, it may be necessary to remove entire stumps. You can accomplish this by a number of methods short of pushing them out with a tractor. Perhaps the simplest way to remove a stump is to use a combination of pulley blocks powered by a hand-operated winch. A sturdy 2-ton winch is the kind of tool put to constant use on a developing homestead. You should consider it one of those essential "first-investment" tool acquisitions, like the chain saw or the mowing machine.

Wherever possible, it is best to terminate tree trunks at ground line and to leave their stumps to decompose in the soil. With some durable species—such as oak, cedar, and locust—stump decomposition takes a long time, and if new growth sprouts, the stump becomes harder and increasingly difficult to remove.

Stumps can be slowly burned out. Drill two 1-inch holes diagonally opposite each other into the stump, and place hot coals where the drilled holes intersect. In a short time the section of stump between the intersecting holes will burn entirely. An oil drum can easily be fashioned into a downdraft stump burner. Remove one end of the drum, place it over the stump, and cut two holes in the opposite end—one for incoming combustion air and one for the emission of smoke. To encourage incineration, first remove the bark from the stump.

Blasting can be the most expedient and economical method of stump removal. Contemplating the use of dynamite is like contemplating the butchering of a hog. Beginning the task seems formidable, and the tendency is to "let the experts do it." But once you butcher your first hog, or once you have set off the first dynamite charge, you will discover that the so-called expertise involved does not warrant outside professional assistance.

Dynamite used to be a standard homestead item, as common as a salt lick or a smokehouse. Every pioneer homesteader handled dynamite for well-digging, road-building, tree-planting, ditching, and stump removal. Powder monkeys—people who use dynamite for hire—can still be found in many rural areas. A greenhorn homesteader would do well to enlist such experienced advice for the first blasting project. In an effort to revive interest in the use of dynamite, we take the time here to acquaint you with some of its significant properties and functions.

Dynamite is sold on the basis of the percentage of its nitroglycerine content. Twenty-percent strength is the lowest charge available and results in a slow, propelling force. Where a rapid, shattering effect is desired, 60-percent strength is required. A slow charge will raise a boulder or stump, while a fast charge will shatter it. A slow charge is set in clay soil, but a fast charge should be used in light, sandy soil. Stump-blasting in sandy soil is not as effective, since the force of the explosion dissipates around the roots.

Generally speaking, 1 pound of dynamite will move 1 cubic yard of earth. You will get the best results when the ground is wet. The secret of using dynamite is in confining the explosive gases completely, for the more resistance offered by soil, the greater will be the force exerted against the stump. For this reason, tamp the explosive in the drill holes thoroughly with a wooden rod.

Use a device called a blasting cap to explode the charge. If you want a single charge, use a fire-ignited fuse to combust the charge in the cap, exploding the dynamite. When a multiple charge is necessary, an electric-detonated cap is preferred. This is not mandatory, however, inasmuch as you may use a series of fire-ignited fuses cut to different lengths for simultaneous detonation.

A blasting machine is used to discharge an electric cap,

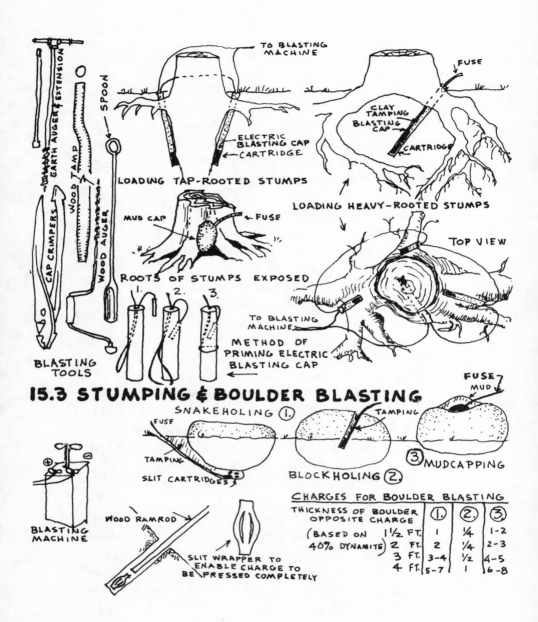

TO BLASTING MACHINE

FUSE

ELECTRIC BLASTING CAP
CARTRIDGE

CLAY TAMPING
BLASTING CAP
CARTRIDGE

EARTH AUGER EXTENSION

WOOD TAMP

SPOON

CAP CRIMPERS

WOOD AUGER

LOADING TAP-ROOTED STUMPS

LOADING HEAVY-ROOTED STUMPS

MUD CAP

FUSE

TOP VIEW

ROOTS OF STUMPS EXPOSED

1. 2. 3.

TO BLASTING MACHINE

METHOD OF PRIMING ELECTRIC BLASTING CAP

BLASTING TOOLS

15.3 STUMPING & BOULDER BLASTING

FUSE
MUD

SNAKEHOLING ①

FUSE

TAMPING

TAMPING

SLIT CARTRIDGES

BLOCKHOLING ②

③ MUDCAPPING

BLASTING MACHINE

WOOD RAMROD

SLIT WRAPPER TO ENABLE CHARGE TO BE PRESSED COMPLETELY

CHARGES FOR BOULDER BLASTING

THICKNESS OF BOULDER OPPOSITE CHARGE	①	②	③
(BASED ON 1½ FT.	1	¼	1-2
40% DYNAMITE) 2 FT.	2	¼	2-3
3 FT.	3-4	½	4-5
4 FT.	5-7	1	6-8

which, like the fuse-ignited cap, must be primed; that is, either kind of cap is set into a stick of dynamite. Make a diagonal hole in the dynamite cartridge so that the cap will lie lengthwise to it. Figure 15.3 depicts some of the features of using explosives for stump and boulder removal.

The best possible use for the commercial fertilizer ammonium nitrate is as a blasting agent. Since 1955, when it was discovered that no. 2 fuel oil could be mixed with ammonium nitrate to make a safe, inexpensive, and readily available explosive, there has been a revival in homesteaders' use of explosives. Add about 1 gallon of oil to a 100-pound bag of fertilizer and thoroughly mix them in a cement mixer. The fuel oil absorbs oxygen when the charge is detonated, confining the charge and thereby affecting the blasting action. Sawdust added to nitroglycerine serves the same purpose as fuel oil in a recipe similar to that just described. Relatively inexpensive fertilizer explosives are best employed on large-scale projects requiring removal of huge stumps or boulders. Placed in large-sized boreholes, regular 60-percent dynamite is used as a primer to detonate the fertilizer. A primer cartridge is usually placed at the top and bottom of the borehole.

After clearing and fencing the homestead boundary line, you must decide where to subdivide the acreage with secondary fences and what fence materials to use for each subdivision. Hard and fast rules are impossible to give, but in general, you should locate permanent fences other than boundary lines along equal contours, at ridge lines, or through draws. A familiar sight in rural America is the badly eroded fence line built from the top to the bottom of a hillside. The continual movement of livestock, which habitually travel to outer pastures along fence lines, causes the erosion of these lines.

It is important to remember that fence-building costs are

TYPE	RELATIVE COST	LIFE (YEARS)	UPKEEP	CATTLE	HOG	SHEEP GOAT	HORSE
BARBED WIRE							
2 PT. 3 ROW 12 GA.	12	40	HIGH	FAIR	POOR	POOR	POOR
2 PT. 4 ROW 12 GA.	13	40	HIGH	GOOD	POOR	FAIR	FAIR
2 PT. 5 ROW 12 GA.	14	40	HIGH	GOOD	POOR	GOOD	GOOD
2 PT. 3 ROW 14 GA.	11	25	HIGH	FAIR	POOR	POOR	POOR
4 PT. 3 ROW 12 GA.	12	40	HIGH	FAIR	POOR	POOR	POOR
4 PT. 4 ROW 12 GA.	13	40	HIGH	GOOD	POOR	FAIR	FAIR
4 PT. 5 ROW 12 GA.	14	40	HIGH	GOOD	POOR	GOOD	GOOD
WOVEN WIRE ☐ = RECOMMENDED					ROWS OF BARBED WIRE NEEDED		
HT. SPACE							
26 12	15	40	MED.	FAIR 4	GOOD 2	GOOD 1	POOR
26" 6	16	40	MED	FAIR 4	GOOD 2	GOOD 1	POOR
32 12	16	40	MED	FAIR 3	GOOD 1	GOOD 1	POOR
32 6	17	40	MED	FAIR 3	EXC 1	GOOD 1	POOR
39 12	16	40	MED	GOOD 2	GOOD	EXC 1	GOOD 2
39 6	18	40	MED	GOOD 2	GOOD	EXC	GOOD 1
47 12	17	40	MED	GOOD 1	GOOD	GOOD	GOOD 1
BOARDS							
1X6	25	50	LOW	EXC	EXC	EXC	EXC
2X6	35	50	LOW	EXC	EXC	EXC	EXC
ELECTRIC							
BARBED WIRE							
1 ROW 12 GA.	4	40	HIGH	FAIR	POOR	POOR	GOOD
2 ROW 12 GA.	5	40	HIGH	GOOD	GOOD	POOR	GOOD
STEEL WIRE							
1 ROW 12 GA.	3	40	MED	FAIR	POOR	POOR	GOOD
2 ROW 12 GA.	4	40	MED	GOOD	GOOD	POOR	GOOD

(Left margin of WOVEN WIRE section: 10 GA. TOP & BOTTOM, 12 GA. FILLER)

15.4 FENCE SELECTION FOR LIVESTOCK

FENCING KNOTS TESTED FOR BREAKING STRENGTH

FIGURE OF 8 – 83-74%   DONALD 73-79%   DOUBLE LOOP   72.27%

relative to the size and shape of the area to be fenced. The larger the area, the less fencing cost per acre. To enclose 1 square acre requires 825 feet of fencing material, but only 33 feet of material per acre are needed when a square section (160 acres) is fenced. A 40-acre field twice as long as it is wide requires 25 percent more fencing material than a square 40-acre parcel.

Contour fencing—curved fence lines following the same

elevation along curving land—is a comparatively new method for defining areas of land use. It is especially useful along terraces, next to diversion ditches, or where plantings are maintained along the contour of the land. You should set posts for a contour fence with the tops leaning out a few inches. When you stretch the wire, the posts will tend to straighten to a plumb position. On curved fencing, you need to stretch the wire to only a moderate tension, much less than is required on straight-line fencing. The sharper the curve, the less tension required. Always remember to place fencing on the outside of posts so that when the wire is stretched it will pull against the posts and not against the staples.

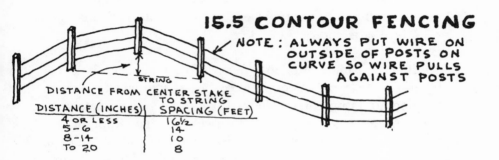

15.5 CONTOUR FENCING

NOTE: ALWAYS PUT WIRE ON OUTSIDE OF POSTS ON CURVE SO WIRE PULLS AGAINST POSTS

STRING

DISTANCE FROM CENTER STAKE TO STRING

DISTANCE (INCHES)	SPACING (FEET)
4 OR LESS	16½
5 – 6	14
8 – 14	10
To 20	8

After you establish a general fencing plan, you must choose fencing materials. Of the various materials for fence-building, line posts represent the greatest cost. Posts can be made of wood, pipe, or cement, or you can purchase commercial steel posts. Choice of post material will be influenced by the following:

1. Availability of raw materials. A well-managed woodlot or a generous source of sand and gravel would influence your choice between wood and concrete posts.

2. Required fence strength. A 1,000-pound force will break a 4-inch wooden post; a 300-pound force will break a 3-inch-square concrete post; a 100-pound force will permanently bend a steel fence post.
3. Soil conditions. Heavier posts are required in sandy soil. For very rocky soil, metal posts may be the only practical choice. Steel posts can be hand-driven into soil of average tilth in 7 minutes, whereas wooden or concrete posts require a setting time of 20 minutes.
4. Accessibility. In places where a pickup or tractor cannot maneuver to distribute fence materials, it is easier to haul a bundle of steel posts weighing 10 pounds apiece than it is to haul a single concrete post weighing 120 pounds. A 20-pound wooden post might in this instance be a satisfactory alternative.

Other considerations may also influence your choice of fence-post material. A concrete post is inexpensive to build, is fireproof and rotproof, and looks substantial and impressive in place. A steel post is ugly and expensive but also fireproof, rotproof, and easy to set. For close confinement of animals (areas such as corrals, loading chutes, or sorting pens) steel posts are inadequate in strength. For temporary, movable fencing or electric fencing steel posts are the best choice. You

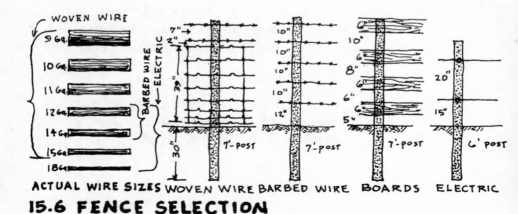

ACTUAL WIRE SIZES   WOVEN WIRE   BARBED WIRE   BOARDS   ELECTRIC

15.6 FENCE SELECTION

can drive them into the ground and later remove them with minimum effort.

An alternative to posts and wire is the "living fence." Besides its effectiveness, a barrier of plants, such as the multiflora rose or Osage orange, has conservation value for the homestead. In some sections of the United States, however— in the warm and wet southland especially—fences made of multiflora rose have invaded fields and become pesty. Be sure to select a plant suitable to your region.

As one aspect of woodlot management, many homesteaders grow wood for posts. Trees of large diameter can readily be split into fence posts. As you split posts, try to retain some heartwood in every one. Such a post will last longer and will hold fence staples more securely. As illustrated in Figure 15.7, first split the log section in half, using steel wedges and a sledge. Then quarter it, as in the sequence shown in the illustration.

If a durable wood such as Osage orange, black locust, redwood, or red cedar is not available, it is necessary to treat posts to discourage wood-rotting organisms. A treated pine post will last up to thirty years, but untreated, it will decay in three years.

You should cut trees for posts in late fall or early winter. Bark peels best in spring and early summer, but rapid summer seasoning results in severe checking (cracking) of the wood, especially in hardwoods like oak. You can control checking by peeling the post as soon as it is cut. Allow the posts to season a few months to remove water, making room for the penetration of preservative. Pile posts in the open so that air will circulate freely around each one.

There are numerous effective methods for preserving wooden fence posts. Which you choose will depend on a wide variety of factors, beginning with the species of wood, the size

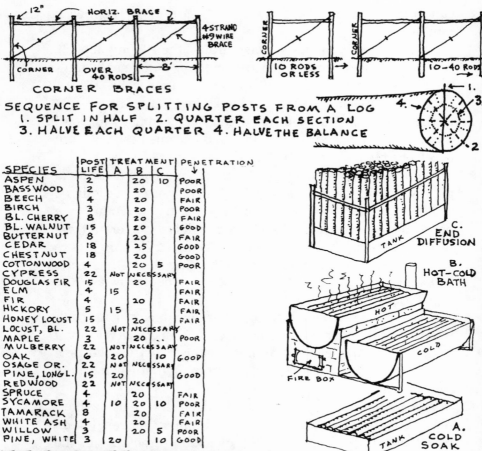

CORNER BRACES

SEQUENCE FOR SPLITTING POSTS FROM A LOG
1. SPLIT IN HALF 2. QUARTER EACH SECTION
3. HALVE EACH QUARTER 4. HALVE THE BALANCE

SPECIES	POST LIFE	TREATMENT A	B	C	PENETRATION
ASPEN	2		20	10	POOR
BASSWOOD	2		20		POOR
BEECH	4		20		FAIR
BIRCH	3		20		POOR
BL. CHERRY	8		20		FAIR
BL. WALNUT	15		20		GOOD
BUTTERNUT	8		20		FAIR
CEDAR	18		25		GOOD
CHESTNUT	18		20		GOOD
COTTONWOOD	4		20	5	POOR
CYPRESS	22	NOT NECESSARY			
DOUGLAS FIR	15		20		FAIR
ELM	4	15			FAIR
FIR	4		20		FAIR
HICKORY	5	15			FAIR
HONEY LOCUST	15		20		FAIR
LOCUST, BL.	22	NOT NECESSARY			
MAPLE	3		20	..	POOR
MULBERRY	22	NOT NECESSARY			
OAK	6	20		10	GOOD
OSAGE OR.	22	NOT NECESSARY			
PINE, LONGL.	15	20			GOOD
REDWOOD	22	NOT NECESSARY			
SPRUCE	4		20		FAIR
SYCAMORE	4	10	20	10	POOR
TAMARACK	8		20		FAIR
WHITE ASH	4		20		FAIR
WILLOW	3		20	5	POOR
PINE, WHITE	3	20		10	GOOD

C. END DIFFUSION

B. HOT-COLD BATH

A. COLD SOAK

15.7 POST TREATMENT

and age (greenness) of the post, the number of posts to be
treated, and the labor and equipment available for this operation. Some methods, it should be noted, will be more effective
than others, providing better penetration of the preservative.
We urge those wishing to investigate the various processes
more fully to review the excellent literature on the subject

distributed by the Forest Products Laboratory of the University of Wisconsin in Madison, Wisconsin. We must necessarily limit our discussion here to two small-scale, low-cost schemes for post preservation: the hot-and-cold bath and the cold-soak methods.

The hot-and-cold bath is almost as effective as treating wood under pressure. Soak wood in heated preservative in an open tank for several hours. Then quickly submerge the post in another tank containing cold preservative for an equal length of time. Heating causes the air in the wood to expand. When cooled, air in the wood contracts, creating a partial vacuum and drawing preservative into wood cells. Keep a coal-tar and creosote solution, mixed half-and-half with used crankcase oil or diesel fuel, at about 200 degrees Fahrenheit in the hot-bath tank. You should maintain the temperature of the cold-bath tank at 100 degrees Fahrenheit, just warm enough to liquefy and thin the oil thoroughly.

A less elaborate method, requiring little work but producing less satisfactory results, is the cold-soak process. Submerge posts for several days in unheated coal tar, creosote, and crankcase oil, proportionally the same solution used in the hot-and-cold-bath treatment. About a half-gallon of preservative, at a cost of 40 cents, is absorbed into each post. The cost can be substantially reduced by the "end diffusion" of posts. With this method you treat only the butt end of each post to a point 6 inches above ground line. Use freshly cut, unpeeled posts. Stand them vertically in a metal trough containing preservative, and allow them to remain there for 1 to 10 days.

In 1917 K. J. T. Eklaw wrote in his book *Farm Concrete* that "ignorance of construction methods and carelessness in handling have been responsible for a great deal of the blame which has been unjustly laid upon the concrete post." This observation is true today. You will certainly fail if you do not observe the essentials of cement construction. The information on concrete-post construction presented below is from our own practical experience.

About eight 7-foot posts can be made from a $3 sack of cement. If a free source of sand and gravel is available on or near the homestead, total material costs, including the necessary steel reinforcing rods, can be nominal. As well as being low in cost, concrete posts are strong, durable, fireproof, and rotproof.

Placing steel reinforcing rods in judicious positions is of great importance in building strong concrete posts. Rods must have adequate tensile strength to equalize the compressive strength of the concrete. Therefore they must be as far as possible from the plane of neutral strain and yet be covered by enough concrete to prevent rusting.

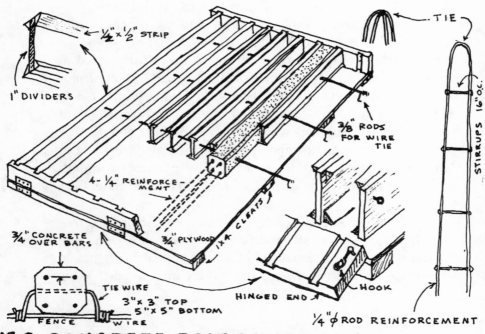

15.8 CONCRETE POST GANG FORM

To begin, bend lengths of ¼-inch rods at their centers and tie them together with wire at the top, strengthening the bond. Before cement is cast in the form, use stirrup-tie wires to circumscribe the bent rods and form a rigid armature.

Using a mechanical vibrator (which may be purchased or rented) will assure a dense, smooth post surface. Otherwise, adequate hand tamping of the concrete in the form may be sufficient to remove air bubbles. A mix of 1 part cement to 5 parts aggregate has proven sufficiently strong, providing the size of the aggregate does not exceed ½ inch. Consistency should be soft enough to permit concrete to settle rapidly in the form when it is tamped. Avoid excessive tamping, which can cause the aggregate to settle out. Remember, excess water weakens concrete. Seven feet is a standard length for concrete posts. The top section should be 3 inches square and the bottom, 5 inches square.

You can build a simple wooden mold, as in Figure 15.8, to accommodate any number of posts in a single pour. The strength of concrete posts is determined, to a large extent, by the curing they receive. You should leave freshly poured posts in the form for several days or until they are thoroughly hardened. Then you may remove the sides and ends of the form to use them on another platform where additional posts may be poured. To cure them, keep the poured posts damp for a week after removing them from the form. Then store the posts in a shady place for several months before using them. Curing posts adequately before using them ensures that they will harden uniformly and it prevents shrinkage cracks. Concrete gains considerable strength for the first year after it is mixed and poured.

For a standard, straight-line, woven-wire fence on level ground, you can space posts 16 feet apart. A barbed-wire fence requires a 12- to 14-foot spacing. Posts for a board fence in an open field are usually 8 feet apart. You should set all posts 30 inches into the ground.

The first step for building any fence is to erect corner posts. About 3,000 pounds of pull is exerted against the posts when a fence is first stretched. A pull of as much as 4,500 pounds can result, for wire will contract in cold weather. Many corner assemblies fail to sustain such force. The corner-brace method shown in Figure 15.7 has proven best, since it offsets the tendency of taut wire to lift anchor posts out of the ground. With this method the pulling pressure of the wire is more evenly distributed among the posts, requiring less effort to construct. Corner posts should be set 3 feet into the ground.

After the line posts are set, stretch the wire from each corner, using double-jack or block-and-tackle fence stretchers. Stretch woven wire until the tension curves in the wire are half eliminated. Do not drive staples so far into posts that the

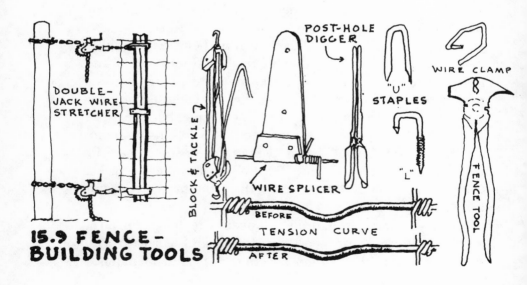

15.9 FENCE-BUILDING TOOLS

DOUBLE-JACK WIRE STRETCHER

BLOCK & TACKLE

POST-HOLE DIGGER

WIRE SPLICER

BEFORE

TENSION CURVE

AFTER

"U" STAPLES

"L"

WIRE CLAMP

FENCE TOOL

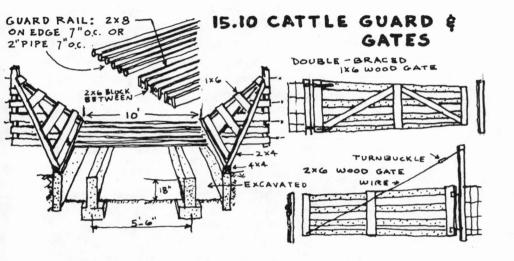

15.10 CATTLE GUARD & GATES

GUARD RAIL: 2×8 ON EDGE 7" O.C. OR 2" PIPE 7" O.C.

2×6 BLOCK BETWEEN

1×6

10'

2×4

4×4

EXCAVATED

8"

5'-6"

DOUBLE-BRACED 1×6 WOOD GATE

TURNBUCKLE

2×6 WOOD GATE WIRE

wire is forced into the wood; wire must be free to expand and contract along its length. Threaded L-shaped staples have holding power greater than the usual wood-splitting variety, especially when used in preservative-treated softwoods.

At various places along fences you must provide for the passage of machinery, livestock, and people. Farm gates should be at least 12 feet wide. A good gate design uses bracing that makes the gate self-supporting. Cattle guards are practical where equipment must pass through a fence many times a day. They consist of a pit 2 feet deep, 6 feet wide, and 10 feet long. You can build railings of steel pipe, cast concrete, or wood planking set on edge. In areas where sheep and goats are turned out, avoid using smooth rungs wider than 2 inches.

You should locate a cattle guard where animals have ample opportunity to study it and evaluate the danger before they attempt to negotiate it. In a farmyard where livestock are held in heavy concentrations, a cattle guard is inappropriate, since animals may inadvertently be crowded onto it. Provide a wide gate adjacent to the guard for herding livestock and maneuvering extrawide equipment. Drawings of practical gates and cattle-guard design are included in Figure 15.10.

SHEEP TRAVELING UP AND DOWN THE SLOPE MADE THE PATHS THAT GRADUALLY DEVELOPED INTO THESE GULLIES.

15.11 FENCE PROBLEMS

ONCE A PROPERTY-LINE FENCE. WATER FROM TWO ADJACENT FARMS EMPTIED INTO THIS GULLY →

16

Roads and Transport

The horse raises what the farmer eats and eats what the farmer raises, but it cannot plow the ground for gasoline.

You do not have to pay a finance company 10 to 15 percent to own a horse.

Horsepower was safer when only the horse had it.   *Will Rogers*

The Volkswagen and the geodesic dome are promoted by their manufacturers as departures from the single-minded conventions of Detroit and suburbia, and as such, many people view them as alternatives. But for the homesteader they are poor choices. The truly viable and creative alternatives are self-styled, site- and climate-attuned housing and reclaimed and refurbished vehicles for work and transport.

The vehicle that meets all our homestead requirements happens to be the one we now drive, a 1954 GMC 6-cylinder, ¾-ton flatbed truck. We chose this vehicle and rebuilt it after thirty years of trial and error with a wide variety of makes and models. This is not to say that our preference excludes consideration of other makes. For many years Ken's personal preference was the 1931 Model A Ford. The 1950s were vintage years for trucks—the decade before Detroit bastardized the pickup, giving it passenger-car roadability by replacing heavy-duty leaf springs with light coil springs. Today, if

16.1 UNIMOG
94 H.P. DIESEL
20 SPEEDS
0.06 TO 45 M.P.H
PRICE: WOW!

MB4/94

money were no object, he would choose the Mercedes-Benz Unimog.

International and GMC are the only two companies that manufacture trucks exclusively. Chevrolet and Ford use the same frame, suspension, and engine in their pickups as they use in their passenger cars. Consequently, even though Chevrolet and GMC are both divisions of General Motors and have many interchangeable engine parts, the older-model GMC proves to have twice the engine performance of the Chevrolet.

The GMC engine has a full, internal, force-fed oil system, not the usual bypass system, along with heavier radiator, piston rods, crankshaft bearings, and cylinder head. For over twenty years the standard GMC 6-cylinder, overhead-valve, low-compression engine remained virtually unchanged. This is an amateur mechanic's dream engine—easy to work on with readily available, inexpensive replacement parts. The overhead "6" offers tremendous lugging power and efficiency, like the rugged, dependable GMC pickup truck.

Older GMC pickups are available and inexpensive. Ours cost $100. We have made a practice of paying a very small first cost for a vehicle. A new ¾-ton truck costs quite a bit more than a ½-ton. But after twenty years there is little price difference, so you are better off with the heavier-duty model. If you had estimated that your homestead needs would require a ¾-ton Chevrolet with a 235-cubic-inch engine, three forward gear speeds, and 15-inch, six-ply tires, consider instead a 1-ton GMC with a 270-cubic-inch engine, four forward gears, and 16-inch, eight-ply tires. The cost for either twenty-year-old truck will be about the same.

The selection process is the same when purchasing a passenger car. After ten years a Cadillac and a Ford have about the same resale value. When both were new, someone paid several thousand dollars more for the better-designed and -equipped, more powerful, safer, and more dependable Cadillac. Old Fords, up to the 1960s, are mostly junk now, but old Cadillacs of the same period can be rebuilt to give thousands of miles of satisfactory transportation.

Once we find the truck we want, we recondition and modify it to fit our particular needs. First, we remove the pickup-style bed; a flatbed with stakes is more practical, especially for light, bulky loads such as manure and mulch. Dual rear wheels and overload springs make it possible to triple the load-carrying capacity. It is easy to install the heavy 1-ton rear axle and springs (found in most junkyards) on a ¾-ton chassis. For optimum performance, it is mandatory to install a two- or three-speed auxiliary transmission, called a brownie.

So equipped, our GMC has fourteen forward gears, and when we leave the flatlands with a 5-ton load destined for our mountain homestead, each gear position is used at least once. Thus, the engine performs at its most efficient "rpm," neither overlugging nor overracing.

When we buy and equip an older truck or other homestead machinery, our second rule is "Overdesign component parts." This makes up for the age and subsequent wear and tear on the vehicle. Most older trucks have a 6-volt electrical system; this should be replaced with a 12-volt system, requiring a new generator, battery, and light bulbs.

Oil cleaners were not standard equipment on older trucks, but if you purchase such a vehicle, it is easy to install a toilet-paper filtration system. Merely replace the filter core with a clean roll of toilet paper plus a quart of oil.

Our GMC train, consisting of a heavy-duty clutch plate, four-speed transmission, three-speed brownie, and twin universal joints connecting an open-line drive shaft to the differential, is extremely rugged and well engineered. Expect only long and satisfactory service from such a unit. Add large (18-inch) dual wheels to the rear and a power winch to the front end of the truck and you almost have a match for the expensive four-wheel-drive vehicle.

Our third rule is "Standardize vehicle make and model." We chose a Chevrolet for a passenger car, since many parts are interchangeable with our GMC truck. By standardizing vehicles it becomes feasible to stock spare parts and invest in specialized engine-rebuilding tools that we can use on both vehicles. Often we purchase spare vehicles for $50, merely to have important parts on hand.

The homesteader of means may purchase a diesel-powered source to meet all equipment needs—the Unimog, which has to be the most versatile four-wheel-drive truck in existence.

Manufactured by Mercedes-Benz, it has twenty forward speeds, ranging from a top speed of 50 miles per hour to 1 mile per hour. The lowest gear is used in conjunction with a front-end bulldozer blade. For traction the Unimog has differential locks on both axles. Detachable equipment includes front-end loader, grader, excavator, compressor, earth drill, loading crane, concrete pump, rotary hoe, snowplow, gravel spreader, and more.

Because of their remarkable lugging ability, diesel engines are the worldwide choice for stationary or mobile work. Essentially, the diesel engine differs from the gasoline, spark-ignited engine in that diesel fuel is meted out in definite quantity to each cylinder for each power stroke according to load. A built-in governor integral with the injector pump controls engine speed. During heavy loading, the diesel engine does not stall, because each cylinder receives an increased volume of fuel via the injector system. A spark-ignited engine slows down during heavy loading, thereby reducing the intake of fuel. This slows the engine further and results in stalling.

As Figure 16.2 illustrates, combustion in a diesel engine is achieved by injecting fuel into compressed air. The compression ratio is 16:1 (16 parts compressed air to 1 part fuel), compared to a ratio of 7:1 for most gasoline engines. This feature gives diesel engines a lower fuel consumption per hour per horsepower. (Considering the lower cost of diesel fuel, one wonders why these engines are not more commonly used in the United States.) Notice the lower exhaust temperature of diesel fuel. The diesel's lower operating temperature and pressure give it a longer life and make repairs less frequent. Because of its relatively clean exhaust, injectors, pistons, and valves are free of carbon sticking.

Diesel engines produce virtually no hydrocarbons; that is,

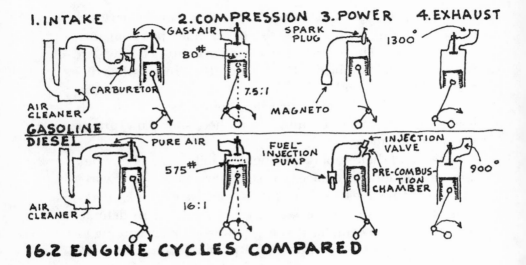

1. INTAKE   2. COMPRESSION   3. POWER   4. EXHAUST

16.2 ENGINE CYCLES COMPARED

they emit no carbon monoxide. When properly maintained, the smoke, odor, and noise produced by a diesel engine are negligible. As air pollution becomes a significant factor, the homesteader might consider converting from a gasoline ignition engine to liquid petroleum (LP). LP is a dry gas that mixes easily with air and burns cleanly and completely, so that it can be used in any high-compression engine. Best of all, it reduces engine wear and upkeep expenses. There is less carbon buildup, less piston ring and cylinder wear, and no crankcase dilution.

Choice of a tractor is a more complex decision than what type of fuel to burn. In many ways a tractor is the heart of the homestead power program—an indispensable tool for clearing land, building roads, leveling sites, terracing, building ponds, and generally maintaining land. A number of factors enter into your selection, including topography, homestead size, soil characteristics, and the scope of homesteading activities you contemplate. Many homesteaders purchase a wheel tractor without considering a crawler-track layer. Before buying a tractor, however, you should be acquainted with the salient features of crawler tractors. We are limiting the discussion here to them because of their safety and suitability to hillside homesteads in particular.

Crawler tractors are ruggedly built. Two structural steel frames connect engine and power train. The power train consists of clutch, transmission, differential, final drive, axle shafts, and brakes. The track frame consists of front idler to keep the upper and lower tracks parallel yet free to oscillate and conform to the ground. Steering clutches transmit power to final spur-type drive gears and thence to the driving sprockets on each axle shaft, which drives the tracks. Steering is accomplished by the tractor driver's using dry-disc, hand-lever steering clutches. When the driver disengages the clutch, power to that particular track is cut off and the tractor turns. For quick, short turns the driver can engage a foot-operated brake.

The above discussion applies to tractors built before the 1960s. Nowadays the tractor industry offers new and sophisticated engine and power-train features. These innovations are attractive in spite of their far greater cost, but they are impossible for the homesteader to repair or rebuild. Hydraulic transmission devices such as the torque converter and fluid drive eliminate manual gear-shifting. These devices utilize the energy of a fluid in motion or under pressure for transmitting power. Thus, the engine meets varying load demands running at its most efficient speed. Some tractors use a two-speed planetary transmission power shift, replacing steering clutches. This makes it possible to shift into lower gear ratios without engaging the clutch. Each track can be driven at a different speed, so that a turn can be made under full power.

16.3 CRAWLER TRACTORS

YR.	MAKE	WEIGHT	FUEL	CY.	BORE & STROKE	H.P	DRAW-BAR PULL-#	RATE
54	OLIVER OC-6	7500	D or G	6	3.4 x 3.7	26	4000	12.14
55	OLIVER OC-12	11700	D or G	6	3.7 x 4.5	45	6500	10.65
55	CAT D2	8500	D	4	4 x 5	34	4100	10.68
56	INTER. TD6	9400	D or G	4	4 x 5.2	38	5300	11.13
56	J.D. 420C	5000	G	2	4.2 x 4	21	2400	8.97
58	OLIVER OC-4	5300	D or G	3	3.5 x 4.5	20	2800	10.61
59	CASE 310-C	6000	G	4	3.4 x 4.2	29	5800	11.30
59	J.D. 4401C	7200	D or G	2	4 x 4.5	25	7000	14.50
59	INTER. T340	6700	G	4	3.2 x 4.1	30	6500	11.60
60	INTER. T5	7000	G	4	3.2 x 4.1	32	7400	12.28
60	INTER. T4	6900	G	4	3.1 x 4	25	7400	12.30
60	INTER. TD5	7100	D	4	3.4 x 4	32	7600	14.00
60	INTER. TD340	7100	D	4	3.7 x 3.8	29	6800	13.10
61	A.C. HD3	7600	D or G	4	3.5 x 4.3	28	8200	12.68
61	J.D. 1010C	7600	D or G	4	3.6 x 3.5	32	7500	13.80
61	CASE 310E	7500	D	4	3.8 x 4.1	34	7000	15.00
62	J.D 2010C	9600	D or G	4	3.8 x 3.5	43	10200	14.11
66	INTER. 500	7900	D or G	4	3.5 x 4	33	7900	13.90

Figure 16.3 itemizes the essential performance features of crawler tractors that are suitable for homestead use. Vintage years for these machines are 1955–65. It would be best to choose your tractor from the "Rate" column of the chart, which refers to the draw-bar horsepower for each hour the tractor is operated and for each gallon of fuel it consumes.

A dozer blade is an essential attachment for any homestead crawler tractor. The size of the blade should match the tractor size, of course, and the size of the tractor should be relative to your homestead needs and soil type. For safety, where the slope of the land exceeds 20 degrees, you should use a wide-track crawler. An angle dozer is desirable for pioneering roads across rough country, and a tilting dozer is handy for sidehill work involving road-building or terracing. You can adjust the blade pitch to ground conditions. The harder the ground, the steeper the blade pitch. Drott Manufacturing Company, a subsidiary of J. I. Case Tractor Company (Racine, WI 53404), makes a front-end tractor shovel that can be used for digging, loading, or transporting material.

Small-acreage homesteaders inevitably face the dilemma of being offered either more tractor than they need or less tractor than they want. The smallest crawler tractor available today may still be too large an investment, yet the garden models offered by most tractor companies are ridiculously inadequate. A good alternative is the Japanese Kubota. Although a wheel tractor, it comes with all-wheel drive, the next best thing to a crawler. Coupled with an eight-speed transmission, the smallest model has a mere 12.5 horsepower, but it is one of the finest two-cylinder, water-cooled diesel engines ever built.

A ruggedly built utility trailer is another important piece of homestead equipment. It can be towed by the crawler tractor over otherwise inaccessible terrain or by the pickup truck on the roads. A trailer can virtually double the loading capacity of a sturdy truck—another good reason for beefing up the power train. Often mobile-home owners discard tandem axle frames after the homes are moved and placed on their site. These frames make sturdy utility trailers.

Finally, a homestead transport rig would not be complete without a front-mounted power winch. The uses of a winch are too numerous to mention—from loading a hayloft to freeing a mired vehicle. Electric winches are the most common, as they are the easiest to install, their motors often being made from aircraft surplus. If you have a choice, however, get the far more powerful hydraulic winch instead. This uses a small hydraulic pump driven by the truck fan belt and connected by hoses to the hydraulic motor in the winch. According to one manufacturer, their hydraulic winch will pull a fully loaded Land Rover with its wheels locked across a dry surface.

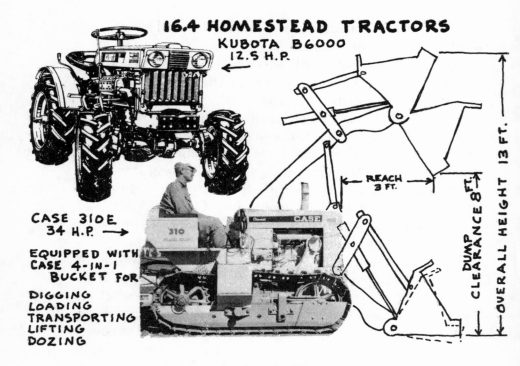

16.4 HOMESTEAD TRACTORS

KUBOTA B6000
12.5 H.P.

CASE 310 E
34 H.P. →

EQUIPPED WITH
CASE 4-IN-1
BUCKET FOR

DIGGING
LOADING
TRANSPORTING
LIFTING
DOZING

REACH
3 FT.

DUMP CLEARANCE 8 FT.

OVERALL HEIGHT 13 FT.

Invariably, the principal development project on any homestead is creating adequate access from a public thoroughfare to homestead buildings. Your truck, trailer, winch, and blade-equipped crawler tractor will all be put to good use developing such a roadway. Late spring is the best time to do road-building—after the last frost but before the ground loses much moisture. Wherever possible, locate the roadway on southern and eastern slopes to minimize the difficulty you will have with snow or ice.

Sound engineering practice, as used in highway construction, should be employed for access roads, too. Stake out the specific alignment for the road. Keep in mind the amount of earth to be filled in and the amount to be cut out so that you can equalize high and low spots. Make certain that earth you deposit in low places is thoroughly compacted. This is another reason for doing the work while the soil is moist. Humus, silt, and clay all lack bearing strength, and you should avoid them as subgrade fill material. Clay, for instance, may be firm when dry, but after a rain it becomes soft and slippery. Subgrade fill material should have good permeability so that water will drain through it readily. Sand and gravel have this quality, but they are difficult to compact. The best subgrade material has portions of clay, silt, sand, and gravel. In the American West, where we live, there is a native material called decomposed granite (DG). It is a rough, coarse sand with enough clay binder to give excellent drainage and compaction characteristics.

Subdrainage is especially important in areas where severe frost damage occurs. You should not use soils that are susceptible to frost, such as fine sands and silts. The proportion of fine materials you use with gravel can be critical. Over 10 percent fines is not suitable, but with less than 5 percent fines the subgrade may loosen in hot, dry weather.

We cannot overemphasize the importance of adequate surface drainage for the maintenance of year-round earth roads. A properly built and maintained road with an aggregate surface can be as serviceable and as long-lasting as the most expensive asphalt road. When soaked with water, however, an aggregate-surfaced road will soften and rut under normal loading. Therefore, you should drain water from the road as quickly as possible. You can accomplish this by crowning the road, that is, maintaining the center higher than either side, and by keeping the surface smooth, firm, and free of loose material. A proper crown may vary from ¼ inch to ½ inch per foot.

You must provide culverts to carry drainage water below

road level. The size of culvert pipe is determined by the maximum amount of rain likely to fall during a given period. You should also consider the rate of flow, the ground slope, and the absorption characteristics.

Homestead roads require periodic maintenance, which is usually accomplished by dragging the surface and depositing loosened material in low spots. You should drag a road as soon after a rain as practicable.

For general road maintenance a simple drag scraper is all you need. The type of drag illustrated in Figure 16.5 was in common use in the Midwest during the 1920s. Similar

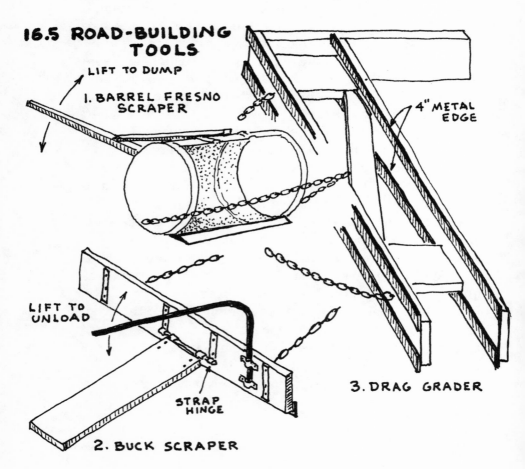

16.5 ROAD-BUILDING TOOLS

LIFT TO DUMP

1. BARREL FRESNO SCRAPER

4" METAL EDGE

LIFT TO UNLOAD

STRAP HINGE

2. BUCK SCRAPER

3. DRAG GRADER

scrapers were used in the original construction of U.S. Highway 1 from Maine to Florida. *The Village Technology Handbook,* published by Volunteers for International Technical Assistance (VITA), gives detailed instructions for building a home-crafted low-cost drag scraper. The VITA handbook also describes how to build a fresno scraper for moving quantities of earth from higher to lower places. One model is made from a discarded oil drum. It is designed so that the power used to fill it also helps to unload it. The buck scraper, also illustrated in the VITA handbook, is used for leveling a road and filling ditches. It is also handy for smoothing out uneven spots created by the fresno. You load the buck scraper by pushing down on the handle as it is dragged forward. Lifting the handle slightly results in a shallow spread of earth. Make the spread deeper by pushing the handle farther forward.

You can easily build all of these implements in the homestead workshop. This is also the place where you maintain the crawler tractor and where you modify and service the pickup truck. As we illustrate in the following chapter, the workshop may be considered the nerve center of any truly productive homestead.

17

Shop and Tools

Every man should have a son and plant a tree.   *Anonymous*

Originating in the Near East, this ancient counsel requires a modern interpretation that would advocate, "Every man and woman should have a child, plant a tree—and build a workshop." Such human activities constitute nothing less than pleasure. And shop work, in which you create new objects and maintain and repair old ones, encourages homestead economy and creativity. The shop is the nerve center of the homestead, the place where you produce and maintain the elements of all operational food-and-shelter systems and where essential tools and implements for the proper functioning of the workshop are used and stored. Therefore, you must carefully consider the location and layout of the shop.

The structure proposed here is specifically for shop work. Its layout was designed to meet shop requirements for an average homestead, with the shop centrally located close to areas of farmstead activity. Considering the hazardous nature of some shop work—which involves welding, grinding, paint and solvent storage, and the use of combustible heating fuel—this building should be prudently constructed of fireproof materials: concrete slab floors, concrete slip-formed walls (see Figure 17.2), and sod (or concrete) roofing. Maximum ventilation is prescribed for similar discretionary reasons. Clerestory windows or skylights eliminate glare from windows and shed indirect, natural illumination, while reserving valuable wall space for tool storage. This building is divided into four work areas, each of which is discussed in detail below.

General Work Space

The general work area should be spacious. To work on the largest homestead equipment—the pickup truck, tractor, and passenger car—a convenient 10-foot-wide, 8-foot-high overhead door is provided. When open, this door occupies no interior working space. A 3-foot-wide swinging door is framed into the overhead door, serving as an entry when the larger unit is closed. This door-within-a-door is specified so that you can carry smaller loads into and out of the shop without the delay and nuisance of having to open the larger door. For personal safety and ease of handling a hand truck or cart, steps into or out of the shop should be avoided. Swinging doors close at the opposite end of the work space, helping to heat the area in winter.

Grease Pit

A good example of multiple uses of space is demonstrated by the design of a pit for the following functions: vehicle maintenance, general parts storage, platform loading, use of a chain hoist on an overhead rail to remove engines and to move heavy objects, and equipment storage when the pit is not in use. A grease pit is indispensable for preventive maintenance on homestead vehicles. The undercarriage of a vehicle must be fully exposed for greasing, oil change, engine

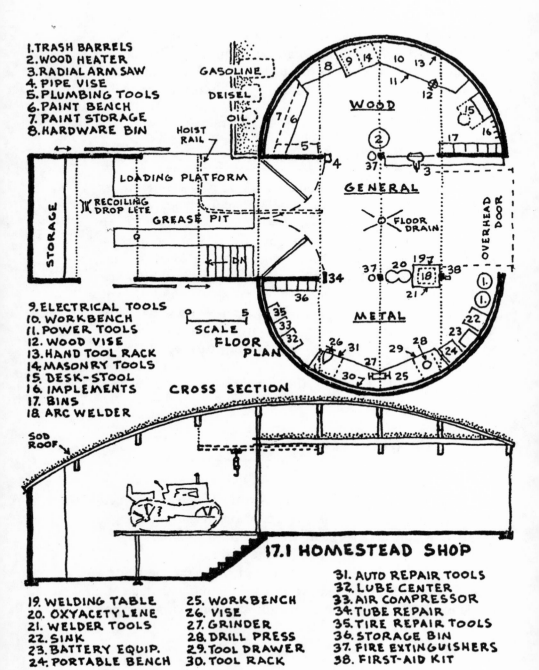

1. TRASH BARRELS
2. WOOD HEATER
3. RADIAL ARM SAW
4. PIPE VISE
5. PLUMBING TOOLS
6. PAINT BENCH
7. PAINT STORAGE
8. HARDWARE BIN

GASOLINE
DIESEL
OIL

HOIST RAIL

LOADING PLATFORM

RECOILING DROP LETE

GREASE PIT

STORAGE

DN

WOOD

GENERAL

FLOOR DRAIN

METAL

OVERHEAD DOOR

9. ELECTRICAL TOOLS
10. WORKBENCH
11. POWER TOOLS
12. WOOD VISE
13. HAND TOOL RACK
14. MASONRY TOOLS
15. DESK-STOOL
16. IMPLEMENTS
17. BINS
18. ARC WELDER

SCALE
FLOOR PLAN

CROSS SECTION

SOD ROOF

17.1 HOMESTEAD SHOP

19. WELDING TABLE
20. OXYACETYLENE
21. WELDER TOOLS
22. SINK
23. BATTERY EQUIP.
24. PORTABLE BENCH

25. WORKBENCH
26. VISE
27. GRINDER
28. DRILL PRESS
29. TOOL DRAWER
30. TOOL RACK

31. AUTO REPAIR TOOLS
32. LUBE CENTER
33. AIR COMPRESSOR
34. TUBE REPAIR
35. TIRE REPAIR TOOLS
36. STORAGE BIN
37. FIRE EXTINGUISHERS
38. FIRST-AID KIT

and transmission removal and repair, and brake and front-end adjustment and replacement.

The left-hand track of this pit is adjustable to fit the varying wheel bases of a truck, auto, or tractor, while the right-hand track, which is part of the loading platform, is stationary. A loading platform makes it possible for one person to move a small hand truck loaded with weighty material whose handling would otherwise require the assistance of several other people. In the shop use leverage whenever possible; you should never lift or carry heavy objects that can be slid along on rollers. A 1-ton chain hoist on an overhead rail is essential for moving extraheavy loads from the loading platform into the shop and for removing and installing engines. To empty the 50-gallon trash barrels (position 1 in Figure 17.1), use the hand truck to move them to the loading platform. One of these barrels may contain burnables; another, scrap metal.

There is a cabinet for general storage at the far end of the grease pit. Store heavier automotive and machinery parts, such as hydraulic jacks and a set of jack stands, at the most accessible level of this cabinet, and stow little-used items at lower- and upper-reach positions. Sliding doors on the left and right sides of the grease pit make it possible to secure the building during the family's absence from the homestead.

Woodworking Space

It is sensible to arrange shop equipment, tools, and materials as close as possible around and in front of the shop worker. Work areas in a semicircular design afford the least amount of movement so that, as in production work, the step beginning a task starts where the last step of the previous sequence ended, eliminating superfluous movement and backtracking. Curving walls support and strengthen the building, giving it an aesthetic appeal seldom found in utilitarian buildings.

The obvious location for the source of shop heat (position 2) is central to the two work areas. Wood heat is specified because wood scraps are produced by, and stored near, the radial arm saw (3). This machine operates best from the position in which it can rip long wood members when both doors at either end of the building are open. These doors are also opened when cutting and threading full lengths of pipe. A pipe vise, preferably a chain type, is located at position 4. It is mounted directly onto a sturdy roof-supporting post for greater stability when strong downward movement is applied for threading or tightening the pipe.

When planning a workshop, you must provide storage for tools and materials near their place of use. There should be a habitual, fixed position for where you will use and store all equipment, tools, and materials. For instance, on a wall panel (5) near the pipe vise you may hang and silhouette hand tools. These may include a three-wheel pipe cutter, a threader of the ratchet type with dies from ½ inch to 2 inches in size, a ratchet reamer, two 24-inch pipe wrenches, and 8-inch and 10-inch pipe wrenches. On a shelf at eye level you can store cans of pipe-joint compound and pipe-cutting oil, measuring tape, crayons, a wire brush, and rags. Below the shelf are a series of bins for assorted fittings, faucets, and odd lengths of pipe. At position 6 is a bench where you can mix paints and other preservative compounds. On the wall above the bench brushes, putty knives, mixing sticks, caulking gun, paint rollers, and paint scrapers are all hung on nails. Open shelves above these tools contain the usual homestead assortment of paints and solvents (7). Putty and glass cutters are also found among the painting paraphernalia.

A floor-to-ceiling bin is furnished (8) for often-used

hardware and electrical items, such as hinges, plugs, and switches. Next to the electrical bin, in the first drawer below the workbench (9), are the electrical working tools: solder gun, acid, solder, plastic tape, wire cutters, screwdrivers, and testers. The workbench itself (10) should be sturdily built and well braced to receive any major impact. Laminated two-by-fours turned on edge make an exceptionally strong bench. On a shelf below the workbench are various power tools (11), including a ⅜-inch portable electrical drill, 6-inch and 8-inch portable circular saws, a portable belt sander, a saber saw, and a chain saw. A wood vise (12) is mounted at the right-hand edge of the workbench, above which woodworking hand tools (13) are silhouetted on racks. We consider the following items to be a minimal collection of tools for homestead wood building projects and general maintenance: two straight claw hammers (13 and 20 ounces), 8-point crosscut saw, 5-point rip saw, 24-inch level, 16-foot box tape, combination square, T-bevel, 30-inch wrecking bar, power bit set, chisel set, slip-joint pliers, screwdriver set, spiral ratchet screwdriver, rafter square, keyhole saw, nail set, gun tacker, 50-foot steel tape, 8-inch block plane, 14-inch jack plane, plumb bob, push-type hand drill, offset ripping chisel, nail puller, nail claw, miter box and backsaw, oilstone, and leather apron.

Another tool drawer is located at position 14 for masonry tools: brick trowel, pointing trowel, two plastering trowels, two finishing trowels, wood float, brick hammer, rubber mallet, hawk, nylon line, line level, and 4-pound stone hammer. Also included in the masonry drawer are tile-working tools: cutter, nippers, and saw-toothed trowel. To minimize weight, the drawers are built narrow and shallow so that they may be removed and transported to projects elsewhere on the homestead. There are guides located inside each drawer so that a sliding tray half the length of the drawer can be installed for storage of smaller items.

At position 15 we have allocated sufficient space for a 2-foot-square, wall-mounted, folding-top desk and a stool. When open, the desk reveals a series of 12-inch-deep shelves designed to hold books, repair manuals, parts manuals, simple drafting tools, and paper. The desk area is also an appropriate location for an intercom for communicating with other centers on the homestead.

The wall panel at position 16 is generally reserved for land-working implements, all of which are silhouetted and hung on nails: hay fork, manure fork, splitting maul, double-bitted ax, pry bar, and pick.

Before returning implements to their respective locations on the wall panel you should clean, repair, and sharpen them. Sharp tools reduce fatigue. Make major repairs, such as replacing broken handles or welding broken parts, during periods of inactivity to avoid delay when you most need the implement. You can recycle used crankcase oil, stored in a 5-gallon drum, by using it as a preservative for wooden tool handles.

You should exercise discretion in the selection of implements. Shovels, for instance, vary in the size of the container, length of handle, and handle-to-blade angle. Square-nosed shovels dig well into even ground, while pointed ones work best in uneven ground. The size of the blade should be relative to the weight of the material you intend to lift. Shovel studies have been made showing that using the proper shovel may increase work output by 20 percent, eliminating worker stress and fatigue.

Stocking tools of only the finest quality also increases work output. Such quality cannot, however, be determined by brand name only. Stanley, for instance, makes fine tools, but they also make a line called Handyman that is pure junk. Avoid combination tools, such as Shopsmith, which compromise the function of one operation to make additional operations possible. Too often the multipurpose tool is too light for heavy work and too heavy for light work. Manufacturers offer thousands of gadgets each year to shop workers, who, like the proverbial fisherman desirous of catching more fish with the latest gimmick, are prone to buy worthless equipment, which only purports to save time and money.

At position 17 we have located compartmentalized floor-to-ceiling bins for storing nails, bolts, nuts, screws, washers, and miscellaneous small items.

Metalworking Space

The arc welder (18) is centrally located in the shop and is indispensable to all metalworking. This piece of equipment, which tucks away neatly under an all-metal welding table (19), replaces a complete blacksmith: forge, bellows, anvil, and countless hand tools. Traditionally, to fuse metals, blacksmiths heated material in a forge and hammered the molten pieces together. This required much patience and skill, for the two pieces had to reach the anvil at the same temperature to make a perfect union. For repairing and maintaining and for constructing new homestead equipment, the arc welder has opened a vast potential for the unskilled metalworker. With about $50 you can purchase a used, 100-ampere, 220-volt AC welder of the transformer type in a size suitable for homestead shop use.

An oxyacetylene outfit (20) is by no means mandatory, but it is certainly handy for cutting and heating metal and welding thin metal. If you cannot acquire used gas tanks with regulators (renting a tank is costly), it would be more economical to rely on an arc welder for heating and cutting metal. The old workhorse of welders is the Lincwelder AC-180. It is manufactured by Lincoln Electric Company of Cleveland and is found in practically every farm shop in the United States. Carbon rods are also used for brazing, soldering, or applying hard-surface powder. Certain kinds of carbon rods are used for cutting metal. Welders other than those made by Lincoln doubtless offer similar design and engineering, but we wager there is no other company in the United States that has produced an article of more practical assistance to the American farmer while at the same time sharing the economic advantages of production with its own workers.

A single metal drawer (21) mounted on the welding table holds welding paraphernalia: welding electrode, gloves, head shield, goggles, chipping hammer, wire brush, and soapstone pencils.

At position 22 a galvanized metal cold-water sink for cooling metal and tempering tools is convenient to the welding center. Its other uses include testing tire tubes, filling batteries, and washing up. A small stand (23) adjacent to the sink holds a few reserve batteries and a battery charger. A 30-ampere-per-hour battery-charging attachment can, incidentally, be plugged into the Lincwelder. On a rack above the charger are hung the few tools for maintaining and installing a battery: hydrometer, carrying strap, terminal cleaner, battery pliers, terminal remover, booster cables, electrical tester, and spare cables and terminals.

A portable workbench with a handy lower tool tray is stored at position 24 when not in use. Since you will most often use it for projects at other places around the shop and homestead, two front wheels should be attached to make it possible to move the bench anywhere.

The metalworking bench (25), sturdily built of laminated two-by-fours, is a facsimile of the woodworking bench and is permanently mounted against the wall. Three essential power tools are secured to this bench: 5-inch machinist vise (26), ½-horsepower bench grinder (27), and ½-inch drill press (28). A narrow drawer below the drill press holds the usual assortment of drill bits, files, punches, and chisels (29).

The rack for metalworking hand tools is placed above the workbench (30), and tools are located at a point on the rack adjacent to where they are most often used on the bench. The most frequently used tools, such as screwdrivers and pliers, are placed within easy reach in a central location. We consider the following tools essential to any homestead project requiring metalwork: 7- and 10-inch locking plier wrenches (Vise Grip), arc-joint pliers, screwdriver set, 12-foot box tape, adjustable wrenches (4-, 8-, and 12-inch), wire cutters, ball peen hammers (8-, 16-, and 24-ounce), 5-pound sledge, hacksaw, tin snips, 10-inch duckbill tin snips, tap and die set (¼-inch to ½-inch by sixteenths, both USS and SAE threads), 24-inch bolt cutters, and "easy out" removers.

In a drawer below the vise are automotive repair tools (31). The upper sliding tray contains a set of ⅜-inch to 1-inch box and open-end wrenches, and a ½-inch drive socket set—⅜-inch to 1-inch with ratchet, drive, and assorted extensions. Below the sliding tray are stored seldom-used engine tune-up and overhaul tools: feeler gauge, valve-seating tool, cylinder-honing tool, valve lifter, piston-ring expander, piston-groove cutter, piston-ring compressor, brake-cylinder ridge reamer, ignition wrench set, and inside and outside micrometer.

Lubricant and lubricating tools are stored at position 32. The different varieties of grease include those specifically formulated for wheel bearings, water pumps, and general bearings. Gear oil and bulk engine oil are stored directly on the floor in 5-gallon containers. Other lubricant equipment includes portable grease gun, hand-pump lubricator for gear oil, spare toilet-paper rolls for automotive oil filters (such as the Frantz filter), various oil cans, and compressed-air grease gun.

This last item is a convenience for the homesteader doing preventive maintenance on a truck, tractor, or passenger vehicle. This tool dispenses grease under the pressure created by a 20-gallon, 1-horsepower portable air compressor (33). A compressor of this size is designed to emit 5 cubic feet of air at 100 pounds per minute, sufficient to operate a grease gun, spark-plug cleaner, blowgun, paint spray gun, engine cleaner, and, of course, a tire pump. There should be an air line with several connections around the perimeter of the shop.

Tire-repair tools are conveniently located on shelves above the air compressor (34). These include the usual assortment of tire irons, mallet, lug wrenches, bumper and hydraulic jacks, portable lever-action bead breaker, hot and cold patches, air-pressure gauge, spare valve cores, and tools. Hot-patch tube repair is done on a wall-mounted clamp located at position 35.

On a homestead numerous small parts for machines are required. Things like pulleys and belts, gears and bearings, spare electric motors and bushings all require some kind of organized storage. Many a trip to town can be averted if it is possible to locate a necessary part when it is needed. A floor-to-ceiling cabinet of individual bins is therefore furnished at position 36.

Finally, at least two fire extinguishers should be visible and readily accessible in central locations, such as at position 37. Be sure you service these devices annually, according to the instructions on the equipment. Do not fail to keep a first-aid

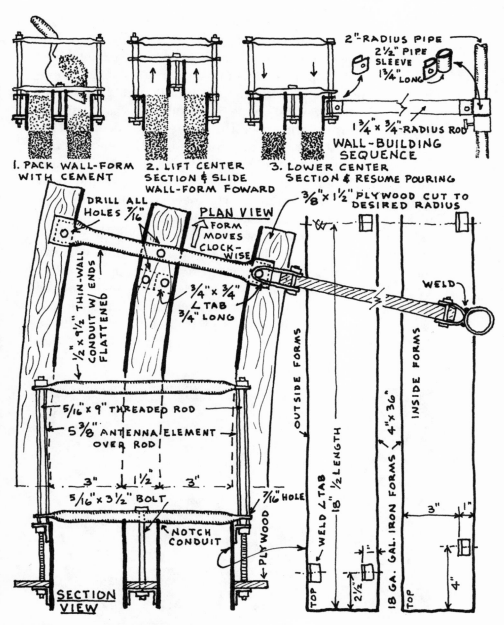

1. PACK WALL-FORM WITH CEMENT

2. LIFT CENTER SECTION & SLIDE WALL-FORM FOWARD

3. LOWER CENTER SECTION & RESUME POURING

2"-RADIUS PIPE

2½" PIPE SLEEVE 1¾" LONG

1¾" x ¾"-RADIUS ROD

WALL-BUILDING SEQUENCE

DRILL ALL HOLES 7/16"

PLAN VIEW

FORM MOVES CLOCK-WISE

3/8" x 1½" PLYWOOD CUT TO DESIRED RADIUS

½ x 9½" THIN-WALL CONDUIT W/ ENDS FLATTENED

¾" x ¾" ∠ TAB ¾" LONG

WELD

5/16" x 9" THREADED ROD

5⅜" ANTENNA ELEMENT OVER ROD

3"   1½"   3"

5/16" x 3½" BOLT

7/16" HOLE

NOTCH CONDUIT

PLYWOOD

SECTION VIEW

OUTSIDE FORMS

INSIDE FORMS

18 GA. GAL. IRON FORMS   4" x 36"

WELD ∠ TAB

18 ½ LENGTH

TOP   2½"   1"

TOP   3"   1"   4"

17.2 SLIP FORM

kit at position 38 or at some similar convenient place where you can readily see and reach it when needed.

Much of the layout for tools and equipment specified here in this model shop existed in our former homestead shop— but, admittedly, not with the organization and efficiency presented here. In our shop the overhead door was too narrow and low, natural lighting poor, the grease pit too deep, and the shop difficult to heat. We accumulated tools over a twenty-year period and too many of them had been borrowed, broken, lost, or misplaced. Sets of things were incomplete and many were obsolete. For instance, who needs a plumber's lead-melting pot in this era of plastic plumbing?

In at least one respect we are envious of young home-steaders just starting to plan and equip a workshop. We envy them the enthusiasm and anticipation of their beginning. They should know, however, that this project will likely be the first they start and the last they complete. Our advice, of course, is to begin by following a plan similar to that outlined above. Good advice also comes from an accomplished shop

17.3 COMPLETED WORKSHOP

user of many years, Milton Wend. In his influential book on post–World War II homesteading he says:

Every activity touched upon requires information, analysis, planning, capital investment, a lot of time and physical effort, and the attainment of that knowledge that comes with actual experience. It is very easy to undertake too many ventures in any one year. It is much sounder practice to draw up a program of proposed undertakings spread out over many years to come. At least there will be no danger of future boredom. Start a few new ones each year if you want to, and prepare for next year's well in advance. The satisfaction due to sound achievement and new skills mastered is more important than the number of activities under way.

18

Animal Products

If we are to master the sensitive arts of building a life-encouraging environment, we need to realize that bigger may not be better, slower may be faster, less may well mean more.   *Stewart Udall*

This precept contrasts sharply with that held by contemporary agribusiness, since the goal of modern animal husbandry is to produce the highest yield of concentrated livestock feed and to feed this material at the fastest rate and in the largest quantity possible to achieve the greatest animal weight gain in the shortest time.

Such methods of animal production have brought grain to the forefront as the ideal, rapid weight builder for animals. To produce a bushel of feed grain, traditional farming methods required 255 minutes, but modern, large-scale, mechanized techniques now produce a bushel in 40 seconds. Yet for some unfathomable reason this fantastic increase in grain production fails to benefit economically the independent livestock producer. On the contrary, feed costs rise as ever more phenomenal feed-grain production records are set. The cost of a 100-pound sack of chicken feed has tripled in the last few years to a price of $9 in 1976, while it still takes about 4.5 pounds of feed to produce a dozen eggs. A hen will probably produce 20 dozen eggs over her sixteen most productive months. The farmer, therefore, invests $9 to feed a $2 pullet for a return of $12 worth of eggs, figured at 60 cents a dozen.

It is not likely that a $1 profit will pay for farm labor, capital investment, insurance, and taxes.

Meat producers experience a cost crunch even greater than the egg producer. Since it takes 10 pounds of grain to produce a pound of meat, there is no way for the perennially debt-ridden rancher to survive the economic squeeze. Ralston Purina Company, the largest feed producer in the United States, will survive, however, as will Tenneco, the immense agribusiness conglomerate.

There is one way for small-acreage homesteaders to rise above the dilemma of high feed costs. They must produce their own feed. For starters, they must reevaluate their whole animal-feeding regimen. As Udall suggests, high yield may be of low value, fast growth may not be as desirable as slow growth, lean may be better than fat. Feed grains containing high concentrations of protein and carbohydrates are not, in the opinion of knowledgeable experts, as valuable as balanced combinations of mineral-rich foraged crops. Furthermore, producing feed grain requires an investment in land and mechanization beyond the means of the average home-steader.

This and the following two chapters give our evaluation of the role animals may play on a well-organized homestead. We have divided the subject of livestock husbandry into products, feed, and shelter.

At the outset you should appreciate the fact that maintaining animals on a homestead is both a luxury and a necessity. It is a necessity if the homestead family intends to consume meat and animal products. As mentioned before, it is undesirable as well as costly to purchase commercially produced animal feed. The abhorrent conditions in which commercial livestock are raised and from which meat and animal products are extracted are unacceptable to us. Animals are force-fed in controlled environments solely for the marketer's economic advantage. They are raised indoors in cages or on concrete or wire mesh. Removed from soil, fresh air, sunshine, and rain, they have little chance for natural exercise lest precious fat or space be lost. Hormones are artificially administered to animals to increase their rate of growth and their body weight. Animals are inoculated with pharmaceuticals so that many of the sickly will not die before slaughter.

Under such living conditions disease is inevitable. Sicknesses, such as Newcastle's disease, which affects chickens; hyperkeratosis, which affects cattle; and bluetongue, which affects sheep, were unknown in 1942 when the Department of Agriculture first compiled its extensive handbook *Keeping Livestock Healthy*. In 1959 an English scientist made a year-long survey of the food value of eggs from chickens raised on free range. Their eggs contained 8,800 units of vitamin A and beta carotene, while eggs from chickens raised in cages had only 4,500 units of these necessary elements. Not long ago, farmers fed chickens 4 pounds of feed for each pound of weight gain. To spur ever greater growth and profit by weight, they reduced this amount by half and replaced the other half with concentrated feed inoculated with antibiotics and dangerous dosages of hormones.

Maintaining animals on the homestead is a luxury as well, for feeding grain to animals to produce meat, milk, and eggs is far less efficient than merely providing the grain directly to people for their consumption. A 100-pound sack of corn has enough food energy (carbohydrates) and cell builders (proteins) to keep a human alive for about sixty days. But, when that same sack of corn is fed to livestock, the animal will produce only enough protein and energy food to sustain a human for five to ten days. WHO published a graphic comparison between the protein requirement for animals and the protein produced by an acre of land (see Figure 18.1).

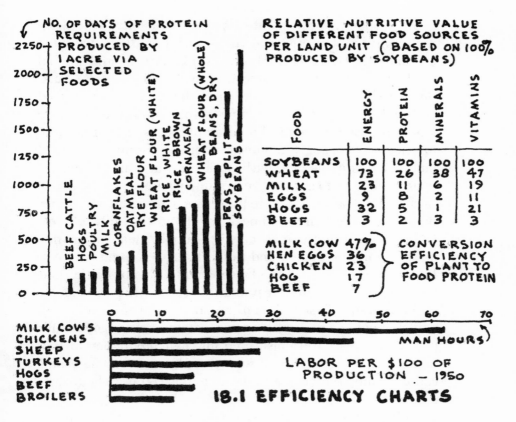

18.1 EFFICIENCY CHARTS

It is a common argument that ruminant animals (cattle, sheep, and goats) do not compete with man for food because their more complex stomach structure digests fibrous foodstuffs that man cannot digest. However, edible pasture and storable feed for animals is grown on land that could be directly utilized by people for their food. The number of domestic animals in the world today roughly equals the human population. As humans increase in numbers, the production of plants supplying high-quality protein, such as oil seed, succulents, and nuts, will compete fiercely with animal production.

Since there is only about 15 percent efficiency when animals convert plant materials into animal products, we must evaluate the number and variety of animals and animal products we require for homestead food production. Beef cattle and sheep are at the bottom of the list of efficient food conversion. Cattle at one time revolutionized agriculture when the males were used for draft purposes. Breeding cattle for meat came later. If a homesteader has abundant low-cost land taxed at a low rate, with sufficient water and a mild climate conducive to growing year-round pasture, he or she may raise a few beef. Cattle require less care than any other livestock. (Maintaining fences is about the only chore.)

Sheep were first domesticated to supply meat. Later they became milk producers and much later were bred for wool. Primitive farmers had a choice among sheep, cattle, and goats. Goats were chosen primarily because they efficiently cleared virgin forested land of brush. They were leaf eaters, while sheep were largely grass eaters. Being good milk producers, goats were also bred for their production of this nourishing food. After land was cleared, however, sheep and cattle became the choice livestock to raise on grassland agriculture.

With the discovery that cattle could be employed to till the soil and give milk, sheep and goats were both less in demand and descended to an inferior position in livestock-breeding, while the dairy cow was elevated to a superior position through specialized breeding. Because of her phenomenal milk production, the dairy cow ranks on a par with poultry in the conversion of plant protein to animal protein. She is twice as efficient as beef cattle and sheep. Pigs rank between milk cows and beef cattle.

Milton Wend is one of the most knowledgeable authors writing on the subject of animals for the productive homestead. He suggests an order of preference for animals as follows: ". . . chickens, a cow or several goats, bees, swine (if one has a cow), a dog, a cat, possibly a horse, rabbits or sheep if they appeal."

We have no quarrel with Wend about the low status he ascribes to dog and cat, for their function on the homestead has been overrated. Cats may rid gardens of gophers, but they kill valuable insect-eating birds, too. Sheep—and we would add beef cattle—are well placed at the end of Wend's sequence. We do, however, take issue with the high place he assigns to bees. Bees' pollination of certain varieties of fruit and nut trees is indispensable, but honey has negligible food value, containing over 80 percent sugar and only minute traces of calcium and iron. A horse, of course, should be stricken from the list. This list was published in 1944, at a time when the horse was still considered viable for draft and transport. Fish must be added to the list.

Our updated sequence must be reordered by some major advances in animal husbandry, by new knowledge of diet and nutrition, by the inflated value of land, by rising land taxes, and by revolutionary progress in plant management. We suggest a homestead animal population of—in order of importance—goats, fish, ducks, and pigs.

The goat has always been, and will probably continue to be, the symbol for subsistence homesteading. It is a maverick, an independent animal, and people who raise goats seem to be of the same ilk. Goats are well suited to marginal-land farming. They thrive on poor pasture, on steep, brush-covered land, and in climatic extremes. They require a certain amount of human attention, which results in a gratifying relationship— one that the agribusiness dairy farmer cannot afford, since it nets no cash return.

In proportion to body weight a goat produces more milk

than a cow, and people can digest goat milk more readily and more easily than cow milk. Although chemically and nutritionally similar to cow milk, goat-milk fat globules are small and tend to be naturally homogenized, making goat milk more digestible for children, the elderly, and those growing numbers who find themselves allergic to cow milk. Cow milk may be suitable for the fast-growing calf, but as pointed out in the next chapter, there is good reason to maintain the human infant's normally slow rate of growth. For this, of course, mother's milk is unsurpassed.

Fish culture is such an important new development that we have reserved a separate chapter to discuss its place on the productive homestead. Where a serious fish-culture program is initiated, a homesteader would do well to include duck and pig populations for balanced fertilization of the pond environment.

Living on the water much of the time, ducks share a natural balance with fish. We know, for example, that they seal a leaky pond with the action of their wings and feet, and they also aerate pond water. Some varieties lay more eggs than chickens, as many as fifty more a year. These eggs are larger (it takes four chicken eggs to equal three duck eggs),

18.2 DUCK BREEDS

WHITE MOSCOVY GOOD GRAZING
POOR LAYERS (45)
GOOD SETTERS
7#

WHITE PEKIN
GOOD TABLE 8#
120 EGGS/YEAR
NERVOUS - POOR SETTERS

WHITE RUNNER
GOOD LAYER (150)
POOR TABLE 5#

KHAKI CAMPBELL
BEST LAYERS (300)
POOR TABLE 5#

AYLESBURY
GOOD TABLE 9#
POOR LAYERS (100)
POOR SETTER

and a duck's productive laying time is two to three times longer. Unlike chickens, ducks do all of their egg-laying at night. Thus, it is possible to allow ducks to range freely during the day without worrying that they will lay their eggs in hard-to-find places. Ducks are like goats in their resistance to disease, while chickens resemble cows in their susceptibility. The death rate among growing ducklings is low, and like goats, ducks thrive on marginal land.

George Orwell was observant when he chose the pig as leader of the farmyard revolution against man. The pig is second only to man in brain power. It is even capable of more original thought than the dog. And of all domesticated animals, the pig is most like the human in its internal physical makeup. A pig has a tooth structure similar to a human's; low-crowned grinding teeth with rounded cusps are suitable for chewing both plant and animal foods. Pigs are naturally clean, being the only domesticated animal that will not soil its litter if allowed outside access. They are lively, intelligent, and extremely adaptable to changes in their environment.

The pig as well as the duck is an important adjunct to a homestead fish-culture program. As we mention in the following chapter, 70 percent of pig dung is food that fish can digest. The labor needed to raise pigs is on a par with that for raising beef cattle and only slightly more than that for raising broilers.

Interestingly, the words for *meat* and *pork* are synonymous in China, perhaps because pork is a complete, easily assimilated food. Pork also contains essential amino acids required for tissue-building. The composition of animal protein more closely resembles the protein of humans than it does that of plants. Animal protein contains a high proportion of the essential amino acid lysine. Vegetables are deficient in lysine; they must be supplemented with animal foods that contain this amino acid. (Such combinations are also mentioned in our chapter on nutrition.)

Domesticated rabbits have always been a source of meat for the small homesteader. About 80 percent of this animal is edible because its fine-grained, highly nutritious meat is about 20 percent protein and 10 percent fat. A doe produces four litters each year, averaging eight to a litter. Fryers dress out to 2 pounds in two months.

The reader may wonder why we have not placed such an efficient meat machine on our selected list of homestead animals. Frankly, no one has yet devised a satisfactory method of producing rabbits except in cages, and we find confining them to wire cages wholly unacceptable. In the following two chapters we discuss ways in which animals may live and feed together for mutual benefit. Rabbits, however, can have no relationship to homestead animal society, because it is necessary to raise them in an unnaturally restrictive environment.

19
Animal Feed

You can breed the pigs and buy the corn and get on.
You can raise the corn and buy the pigs and get on.
If you buy the corn and buy the pigs to feed, you haven't a chance.
But if you breed the pigs and raise the corn, you'll make money.

Louis Bromfield

       Heedless of persuasive arguments like Bromfield's, many farmers since the 1920s have veered from raising animals on free range and putting up silage and hay (raising the pig and the corn) to producing animals in feed lots (buying the pig and the corn)—which is tantamount to factory farming. Livestock management of this magnitude has propelled commercial feed manufacture to ninth place among the foremost industries of the United States, giving rise to such giants as Ralston-Purina. (This growth parallels the enormous expansion of the human-feed industry, with its processed and packaged "food products.") One-sixth of America's six thousand manufacturers of animal feed account for 80 percent of its production.

The homesteader has less reason than the small farmer to buy animal feed. We feel that if homesteaders cannot supply feed from their own place, they have no business keeping animals. Economic considerations only partly support this contention. Primarily, we consider commercial feeds unsuitable for producing healthy animals or nutritious animal products.

The modern concept of animal, as well as human, nutrition

is based, for the most part, on the value ascribed to the caloric and protein content of foods. Evaluated on this basis, the resulting fast growth of livestock—and children—is deemed highly desirable. Yet, with children in whom such rapid growth has been stimulated, we find that their bones age prematurely. This focus on concentrated, tissue-building, fat-riddled foods has created obese creatures. Not only is one out of five Americans overweight, but most have lost sight of what constitutes natural growth in ourselves and in our livestock.

The ratio of nonessential to essential fats in domesticated animals is 50 to 1. In wild animals it is 2 to 1. Essential fats, called lipids, help form new cell structures. The earliest animal growth takes place in the embryo stage, during which the nervous system develops faster than the remainder of the body. After birth, bone structure assumes the more rapid growth. Next, the young animal develops lean musculature. Fat deposition should occur only after an animal reaches maturity.

A diet of commercial animal feed is far too rich in digestible protein. Mineral-deficiency diseases are apt to occur when these saturated rations are fed to young livestock. And being high in carbohydrates, commercial feed creates excessive fat in animal tissue. Barley rations are fed, for instance, in concentrated amounts a few weeks before butchering in order to "finish," or distribute extra fat throughout, the meat. This makes for marbled flesh, undesirable for human consumption. Compared to the animal pastured on free range, animals fed commercial feed live barely a full lifetime. For the most part they are the sickly victims of obesity that must be marketed before death overtakes them.

Some animal nutritionists claim that 50 to 90 percent of a pig's food requirement can be met with proper pasture

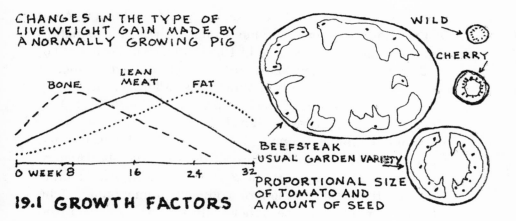

CHANGES IN THE TYPE OF
LIVEWEIGHT GAIN MADE BY
A NORMALLY GROWING PIG

BONE   LEAN MEAT   FAT

0 WEEK 8   16   24   32

WILD
CHERRY
BEEFSTEAK
USUAL GARDEN VARIETY

PROPORTIONAL SIZE
OF TOMATO AND
AMOUNT OF SEED

19.1 GROWTH FACTORS

management. The normal changes occurring in the live-weight gain of a growing pig are, for example, illustrated in Figure 19.1. Unlike ruminants, however, the pig has a relatively small number of cellulose-splitting bacteria in its digestive tract. A pig's diet, therefore, should be limited to 5 percent fiber. The diet for cattle may be as much as 50 percent fiber. Perennial comfrey or a succulent fast-growing annual, such as rape, makes excellent pasturage for pigs. High in water content, succulents are an important addition to the diet of all animals, since water is essential for the transport of nutrients within the animal body. Water also plays an important part in body metabolism and should be available to animals at all times. Also excellent for pasturage are the nitrogen-bearing legumes, such as alfalfa, ladino clover, field peas, soybeans, and vetch.

In a previous chapter on sod crops we mentioned the Swiss dairymen who doubled the protein value of animal forage by harvesting grass at an immature stage. When grass is harvested or grazed at what is called the jointing stage, the percentage of nutrient increase is even higher. Dr. C. F.

Schnable, a biochemist from Kansas City, found that 1 pound of jointing grass contains more vitamins than a ton of mature grass. It was Schnable who in 1930 found that the amount of nutrients in grass reaches a peak just before the reproductive stage, on the day the first joint starts to form. Fully grown oats, for example, are 5 percent protein, but in their jointing stage the amount runs as high as 45 percent.

Plants combine amino acids, creating for themselves usable protein structures. When consuming vegetable protein, animals digest these structures, again creating simple units that are then realigned to form complete animal protein. One kind of protein may be deficient in one or another of the amino acids, but if two foods with complementary amino acids are eaten together, a whole or complete protein is formed. (We also mention protein combinations in a later chapter on nutrition.)

A pea-bean meal can be an excellent source of homegrown protein for livestock. You should cut the plants when the stems and leaves are green and tender—in the stage, as noted previously, when they contain the highest amount of protein. Grind the well-cured (dried) plants in a hammer mill, an essential tool for processing many homestead foodstuffs. You can grind cob meal, corn kernels and cobs, for chicken feed; it is only 65 percent as nutritious as pure cornmeal, but it offers roughage. You can also feed cob meal to rabbits, although feeding only grain to animals is a questionable practice. Rabbits need large amounts of green feed. For complete digestion of ground meals or grains, supplement these coarse foods with root crops from the garden and with alfalfa or clover hay from the pasture.

Diversity in diet is as important as regular feeding. We know now that certain foods release in other foods nutrients that would otherwise be lost. A single-food diet, high in energy-producing carbohydrates, is fed to animals only at the expense of their need for minerals and oil-soluble vitamins. You should feed an entire plant to an animal (as opposed to just the leaves) because the minerals in seeds, stems, and even roots are valuable to an animal. Oil-rich seeds provide the medium for the transport of soluble vitamins to cells. On one hand plant propagationists scurry to develop new strains of high-yield hybrid (seedless) plants, while on the other hand nutritionists hasten to formulate supplements for components found naturally in seeds. Ever hear of the seedless tomato? To get full nutritional value from this wonder variety (see Figure 19.1), it should be eaten with supplemental vitamin C to replace the natural C usually contained in its seeds! Another improvement tragedy happened in 1970 when 700 million bushels of corn from the Midwest were destroyed by a virus that attacked the leaves and tissues of sterile male corn—that is, a hybrid variety of corn. Open pollinated varieties were not touched by the blight.

The frequency of animal feedings greatly influences how efficiently an animal assimilates nutrients. A goat requires twice as much food per gallon of milk produced if the feed is supplied at a single feeding rather than at five. Animals cannot sufficiently digest nutrients when they are packed into the stomach all at once for long periods of time; therefore, it is best to feed little and often. Better yet, let animals graze, choosing their own diet. As illustrated in the valuable manual *Observations on the Goat,* published by FAO, an equation for animal nutrition requirements will read:

$$R = aM + bG + cL + dF + eW.$$

This translates to

Requirement = Maintenance + Growth + Lactation need + Fetal development + Energy for movement.

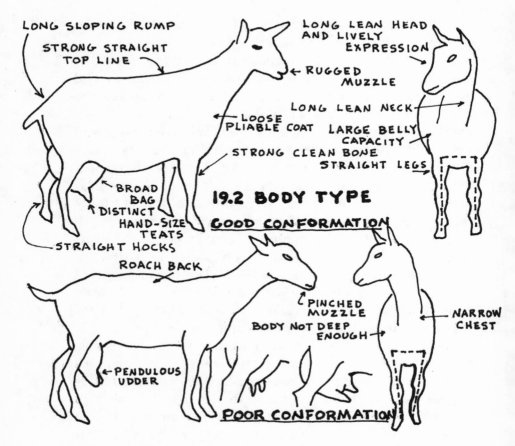

LONG SLOPING RUMP

STRONG STRAIGHT TOP LINE

LONG LEAN HEAD AND LIVELY EXPRESSION

← RUGGED MUZZLE

LONG LEAN NECK

LOOSE PLIABLE COAT

LARGE BELLY CAPACITY

STRONG CLEAN BONE

STRAIGHT LEGS

← BROAD BAG

↑ DISTINCT HAND-SIZE TEATS

19.2 BODY TYPE

GOOD CONFORMATION

STRAIGHT HOCKS

ROACH BACK

PINCHED MUZZLE

BODY NOT DEEP ENOUGH →

NARROW CHEST

← PENDULOUS UDDER

POOR CONFORMATION

Interpreted by feed-lot practitioners, *requirement* means something quite different from the homesteader's understanding of an animal's nutrition requirement. It is possible that it is physiologically advantageous to be hungry part of the time, to strive for food. Nutritionists experimenting with rats found that when they reduced the rats' food supply, their life span was increased 30 percent over that of rats fed in the manner of feed-lot cattle. Underfed goats produce milk that is rich in butterfat, while overfed goats produce milk in greater quantity but poor in butter-fat.

A well-organized homestead provides animals with controlled access to garden and orchard. Pigs and sheep pasture well in a harvested potato field. Chickens and ducks do well in orchard and berry patch, destroying insects in the process of their foraging. Ducks are scavengers. They thrive on parasites, such as husk or liver fluke, which ordinarily kill sheep, cattle, and even chickens. It can be advantageous to have goats graze in a mature orchard, provided that some sort of halter is used to prevent them from reaching into the branches of trees.

Whenever possible allow animals to graze together or follow one another. Ducks will follow pigs, and pigs will follow cattle. Cattle, sheep, and goats graze together companionably because each animal prefers different proportions of leaves and stems in their diet. Cows like tall grass, sheep relish short grass, and goats thrive contentedly on woody brush. Having a split lip, sheep are able to bite closer to the ground than cattle. They graze less uniformly, however, than cattle do.

Contrary to popular opinion about grazing, a pasture in good condition need not be fallowed, plowed, or reseeded. The presence of weed plants in a pasture is indicative of productive rather than unproductive ("poor") soil. Weedy hay from such a pasture has a better balance of minerals than seed hay from fertile soil. The practice of using additives high in nitrogen to force pasture yield is likely to result in a lower protein content of the crop, especially in its mature stage.

It is possible for young pasture grass to be too rich in nutrients yet too poor in minerals to be satisfactory for grazing animals, especially young ones. It has been found that pasture plants grown in dry regions have a higher nutritive value than the same plant grown in a climate of above-average rainfall. Pasture growth is increased by rapid recycling of nutrients. Grazing animals do this better than tillage. Decomposed plants from animal digestive tracts release minerals to the environment faster than vegetative decomposition.

Seeding may even be done by animals. Hard-shelled legumes fed to animals will be spread throughout the pasture in the animals' droppings, reseeding the area—provided, of course, that the soil contains adequate humus. Harold Heady, professor of forestry at the University of California, conducted a study of the effect of spreading humus-producing mulch on annual grassland. When mulch was removed before the new growing season, a reduction in the growth of grass occurred the following year. An average of only 1,000 pounds per acre of new growth was realized when all mulch was removed, in contrast to a yield of 2,300 pounds per acre when mulch was left on the land.

It would be erroneous and a disservice to the serious reader to leave the impression that almost any soil-depleted, drought-ridden pasture could provide feed for any number of animals. Pasture-grazing requires close supervision and control. You must plan cross-fencing so that the size and shape of the field are relative to its soil conditions, existing vegetation, wind, and topography. Fencing makes it possible to plan field rotation designed to keep all herbage evenly grazed. Livestock tends to congregate near water and a salt lick, so you should locate these amenities on opposite sides of a pasture to promote more even grazing. You may have to build trails and roads to less accessible areas of pasture. Cattle prefer level land and avoid grazing steep slopes. Sheep customarily face into the wind while grazing and tend to overgraze the windward side of a field. Knowing the direction of prevailing winds is therefore necessary before you install cross-fencing.

With proper planning, even in areas with short growing seasons, you can maintain the heavy-feeding milk cow entirely from homestead-produced foodstuffs. One acre of pea-bean meal or cob meal, supplemented with carefully apportioned root crops from the garden, will keep a cow in good milk production throughout the year.

On free range the pig may fare even better than the cow. There is an old farm joke about the adaptability of the self-sufficient porker whose freedom to run about and feed freely was criticized by its owner's neighbor. "If you fed that pig in a pen, you'd get it to market in half the time," the neighbor declared. "Mebbe so," the farmer replied, "but what's time to a pig?"

Time does have a way of catching up with all people and all seasons—and certainly with the insatiable pig. Eventually free range and fresh young grass give way to winter dormancy. It is then that you must feed domesticated stock a ration prepared earlier when plants were at their optimum maturity. You should store this silage in a manner that will preserve its nutrient value.

The first agriculturist to reconcile the nutrient losses customarily experienced with haymaking was a French farmer, Auguste Goffart. In 1877 he built the first silo and described

19.3 MILKING PRACTICES

DOOR ←16'→ 39'

PLAN

19.4 GOFFART'S SILOS   1877

its construction in his *Manual of the Culture and Soiling of Maize and Other Green Crops.* In northern Europe, where unpredictable summer rainfall and low temperatures made it difficult to cure hay properly, Goffart solved the problem of making hay. Known as the father of modern silage, he raised enough feed to keep 100 head of cattle year-round on his 86-acre farm in Burtin, France.

Goffart's silo (see Figure 19.4) was a shallow structure 16 feet wide by 39 feet long and 16 feet high, with a door opening at one end. Goffart thoroughly understood the principle of controlled fermentation. He knew, first of all, that the volume of air in silage must be controlled. This he accomplished by placing boards on top of the silage, after which he weighted down the mass with a foot-thick layer of earth. He also understood that it was essential for a certain degree of acidity to develop in silage as quickly as possible. When the living crop is cut, chemical changes continue by plant respiration and by bacterial enzyme activity. An acid condition controls these changes, which would otherwise putrefy the mass.

Goffart later found that the weighted boards and earth could be replaced by the use of a taller structure with a smaller diameter. In a towerlike container, less silage surface would be exposed to air than in a shallow container. Silage in a deep container would also be more thoroughly compacted by the weight of the material above.

Immediately after Goffart published his methods for ensiling fodder crops, farmers in Europe and America set to work building silos of all shapes and descriptions. The first were square towers built of wood, but friction prevented thorough compaction of silage in corner spaces. So octagonal and, later, circular designs were adopted. But wood was not the ideal material, because repeated shrinking and swelling destroyed its elasticity. After several seasons' use the structure would not return to its original shape.

Farmers eventually resorted to masonry materials for silo-building, including brick, block, stone, and poured concrete. Cast concrete proved to be the ideal material because it is fireproof, durable, wind-resistant, and inexpensive. In the early days of silo-building great care had to be taken when casting concrete walls, to insure against leakage. Silage juices, containing lactic acid, can readily seep into silo walls and dissolve the cement binder. Today we use plastic materials to coat concrete walls, protecting them from dissolution.

Not many years after Goffart wrote his thesis, F. H. King, professor of agriculture at the University of Wisconsin, formalized the process of making silage by lending it scientific credence. Much of what we know today about silage production comes from the disciplined research of King. He also provided us with some highly imaginative silo designs, which will be described and illustrated in the next chapter.

In 1900 King stated that a minimum of eight cows (or their equivalent) were required to justify the use of a silo. He based his estimate on the fact that from 2 to 3 inches of silage must be removed each day from an 8-foot-diameter silo to prevent mold from forming. The warmer the day, the more silage that must be removed to prevent spoilage.

Since King's time the efficient slip-form technique for building small-diameter concrete silos has been perfected, making it economically feasible to reduce silo capacity by half. A single cow (or two goats) can consume 40 pounds of silage a day. Over a six-month fall-winter feeding period, this amounts to 7,200 pounds of silage, requiring 240 cubic feet of silo space. To this quantity we must add the 10 pounds consumed each day by a single hog and the variable amounts eaten by chickens, ducks, and perhaps a few beef cattle. Allowing for a percentage of waste, a silo with a minimum 15-ton capacity is required. A silo of this size, 20 feet high, is the most efficient to build, fill, and maintain, and it is a reasonable size for the average homestead. Four cows can be supported from a silo of this capacity, described in detail in the next chapter.

A homesteader should allocate 1 acre of land to stock a 15-ton silo. While 1 acre of hay yields only 2 tons, 1 acre of fodder corn will yield from 40 to 75 tons. Obviously, considering the amount of nutrient received per acre of land farmed, a hay crop is the more expensive choice for animal feed. As noted before, the food will lose considerable value between the time of cutting the hay crop and actually feeding it to livestock, but well-made silage has approximately the same food value as the original plant. Consequently, dairy farmers find it a suitable substitute for grain feed.

What you choose as a silage crop depends on many factors, such as soil, climate, the variety of animals to be fed, and the degree of mechanization at hand. (Making silage requires far more equipment but less labor than making hay.) In any case a silage crop must be a mixture of plants rich in fermentable sugar and high in protein. The proper mixture has a full complement of protein balanced with other nutrients and is suitable for feeding to all homestead livestock. Such a balanced mixture would contain, for instance, alfalfa hay, which is low in energy value but has twice the protein of corn. An ideal mixture is one with two parts alfalfa to one part corn. Farmers who feed grain to livestock place far too much emphasis on protein content and too little on the energy requirement for the animal. Soybeans by themselves are too high in protein. Rather you should offer a mixture of one part soybeans to four parts corn. Sorghum is a first-choice crop for silage in various parts of the United States. Many farmers prefer it over corn, but like corn it contains an excess of fermentable carbohydrates. A leguminous crop, such as cowpeas, will balance the mix. You can grow and harvest cowpeas and corn in alternating rows, making excellent silage for hogs. Clover and corn are excellent winter feed for poultry.

About any imaginable combination of plants can be successfully made into silage. Farmers in Europe prefer to make silage from garden and root crops, such as beets, beans, steamed potatoes, sunflower heads, turnips and cabbages (but go easy on these gaseous vegetables), and windfall fruit. These succulent crops act as preservatives for the very young chopped grass and leguminous plants that are added to the mix.

Some unusual combinations of plants have been made into silage. In Arizona mineral-rich spineless cactus run through a feed cutter produced excellent silage feed for one farmer. Russian and Canadian thistles become soft, palatable, and extremely nutritious after two weeks in a silo. Weeds, leaves, ferns, and even twigs can be ensiled.

Producing good silage is like making good wine—many of the same principles are employed. As discussed before, the blend must be proportionally correct. But even before this, it is of utmost importance that you use plants in their optimum stage of growth. With grass this is the late leaf stage. (see Chapter Ten). In general, these grasses should be cut after the leaves have unfolded and after the heads (buds) have emerged but have not bloomed. Exceptions to this are alfalfa, which may be mowed at a stage of early bloom, and red clover, which should be cut at the mid- to full-bloom stage. Soybeans and cowpeas may be mowed when the first seed pods are filled.

As a grass plant matures, its protein, mineral, and vitamin content decreases, while crude fiber is increased and digestibility is reduced. Perceptive farmers in the Middle West use a rule of thumb that states that for every day's delay in harvesting a grass crop after June 1, its value as silage decreases 1 percent.

Fermentation, essentially, is a method of preserving foodstuffs. When making good wine or good silage, the goal is to control the process of decay. This is accomplished in two ways: by encouraging an environment conducive to the growth of anaerobic bacteria, which quickly create the acidity that delays ongoing chemical changes in newly harvested plants; and by minimizing the volume of air in the mass, since air and excessive moisture are the chief impediments to good silage- (or wine-) making. Even after plants are cut, oxidation (decay) will continue apace in the still-living organic matter unless its condition is altered.

The first concern in silage production, then, is to create a certain acidity in the mass. This is done by enlisting the aid of hyperactive anaerobic bacteria, which thrive excitedly in this medium; that is, they live in "ferment" in the oxygen-poor environment. Their enzymes, in effect, rush to convert plant starches to sugar and from sugar to lactic acid and carbon dioxide. Plant respiration is slowed significantly as soon as the pH level descends to 4.0. Ideally, this pH level is reached soon after the silo is loaded. At this stage aerobic microorganisms are still alive but cannot grow in the high acidity any more than they can grow without oxygen.

Oxygen is essential to the decomposition of organic matter. The best way, therefore, to preserve any crop is to replace its oxygen with carbon dioxide. Oxygen is regulated by displacing air, and this is accomplished by holding the en-siled mass in a towerlike container that, by the nature of its shape, compacts silage and seals itself by the sheer weight of the material in the stack. The weight of silage creates a pressure on silo walls on the order of 11 pounds per square foot of wall area per foot of depth. For our proposed silo 20 feet high, the lateral pressure would be 220 pounds per square foot.

Only the top surface of the material in the silo is exposed to continuing oxidation. A level surface, which exposes a minimum of the mass to air, is maintained by the daily removal of the top 2 or 3 inches of decomposing silage. The main part of the crop is preserved, however, as long as anaerobic activity is maintained in an airless environment.

A homestead-built concrete silo should be tightly packed, particularly around its edges. Chopped fodder packs faster and more thoroughly, and it also releases plant sap, which aids in fermentation. Therefore, fodder should be cut and chopped in one operation. If it contains extra moisture, it will, of course, weigh more and be more firmly compacted and self-sealing. The moister silage is, the easier it becomes to regulate oxygen by air displacement. Too much moisture, however, causes seepage and nutrient loss.

To control the moisture content of silage, King originated the practice of wilting the crop after it is cut, while it is still in the field. The percentage of dry matter present in grass when it is ensiled is a factor of considerable importance. Wilting grass dispenses with the need for the preservatives so widely used to produce commercial silage hastily. Active fermentation begins as soon as the silo is filled with slightly wilted fodder, about 24 percent dry matter. As stated above, the rapid formation of lactic acid is of utmost importance in this process. If the fodder is too wet, fermentation begins slowly and may last up to 6 weeks. If, on the other hand, the silage is more than 35 percent dry matter, a rise in temperature will occur, destroying crude protein and making spontaneous heating of the material inevitable. The result is a dangerously flammable situation. Digestible crude protein in silage varies from 0.5 percent in wet grass to 28 percent in wilted grass.

One of the many pleasant tasks of making silage is squeezing a handful of the fragrant green to determine the proper stage of wilt in chopped fodder. When the level of moisture is just right, the tightly squeezed fodder will recover its original volume when released. If it expands suddenly and then disintegrates, the crop is too dry. If it remains in a tight ball and exudes moisture, the fodder is too moist. In any event, the sweetness of the aroma of new hay is but one of many recurring compensations to be garnered from the simple tasks of homestead life.

20

Animal Shelter

The biotic power of an animal is limited only by the repressive forces in the environment.   *Samuel Brody*

Having given you fair indication of how feeding affects animal growth and productivity, we now direct our attention to the environment in which domesticated animals must live. In the final analysis the homesteader's ability to create and maintain favorable living conditions for each species brought to the homestead determines its productivity.

Once again, we begin a chapter by describing some of the work done by our mentor, F. H. King. Primarily the foremost soil scientist of his day, King also did much original thinking about the environmental needs of livestock. In the 1907 edition of his *Textbook of the Physics of Agriculture* he proposed some totally unorthodox concepts for animal housing. Barns, he maintained, should be consolidated, bringing all animals together under one roof. One of his barn designs was circular with a central, cylindrical silo. King claimed that this structure was the least costly, most labor-efficient, and structurally soundest barn a farmer could build.

Countless barns have been erected in the intervening years since King's book first appeared, but few incorporate his proposals. To this day barns represent the personal pride of their farmer-builders, yet most provide a low level of function for the labor and the cost of erecting them. For example, a dairy building costing only $200 per cow to construct will produce milk of quality equal to that produced in a building costing $500 per cow.

Even barns designed expressly for homesteaders by people who purport to know about animal housing needs are notable only for their badly planned features. Ed Robinson's small "Have More Plan" barn is a structure 16 feet by 30 feet that houses thirty hens; varying numbers of broilers, fryers, and pullets; six pens of rabbits (which never see sunlight); four goats; and three pairs of breeding pigeons. In plan the barn is a maze of pens, stalls, roosts, nests, and places for feed cans, milk cans, and water cans. Some may call this generous use of partitions and doors "organization," but any animal housed

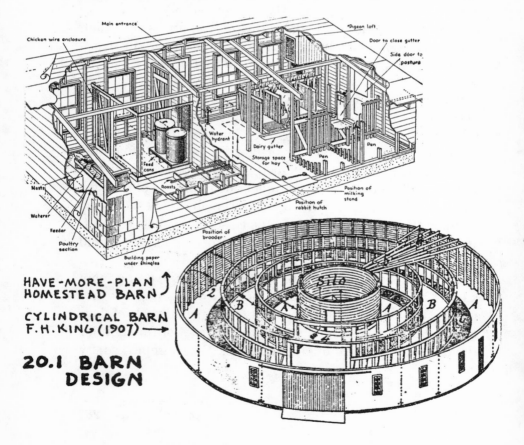

HAVE-MORE-PLAN HOMESTEAD BARN ↑

CYLINDRICAL BARN F.H. KING (1907) →

20.1 BARN DESIGN

in this manner must experience a feeling not unlike that experienced by people inhabiting the typical five-room box house in stifling suburbia.

The first principle of good barn design is flexibility of space. As soon as a pen, roost, or nest is nailed down, that space becomes dead for future expansion. While a barn does often have to provide special facilities, they should be adjustable and movable. Partitions may be necessary at times, but when not in use they should slide or fold out of the way. Homestead animal populations are in constant flux. At times, extra space may be needed for the newly born, whereas a day's butchering activity may leave a barn half empty.

We are partial to King's suggestion for a centrally located silo where silage satisfactory to all animals (as discussed in the previous chapter) can be stored. In one of King's circular barns (see Figure 20.2) you can bring hay into the barn on hayforks through a dormer. The hayforks ride on a track up the curvature of the barn roof, dropping hay on the center of a hay tipple. The tipple, mounted on a rail, runs completely around the silo, dumping hay wherever it is needed in the loft. A similar system of carrier tracks distributes silage and removes manure. Instead of the contemporary blower pipe, a silage carrier fills the silo.

Centrally located, the cast-concrete silo is important as structural support for the barn. King makes the point that the shape of his cylindrical barn resists excessive wind pressure. Openings in barn walls should be of sufficient size to allow machinery (pickup, trailer, or tractor) to pass through. These openings, built of two thicknesses of fiber glass, should be several in number and supply summer cross-ventilation, light, and access for humans, animals, and equipment. This three-in-one door obviates the need for separate window installations. Instead of south-oriented solar fenestration, such

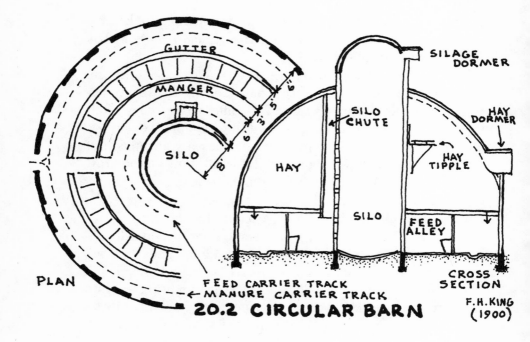

GUTTER

MANGER

SILO

SILAGE DORMER

SILO CHUTE

HAY DORMER

HAY

HAY TIPPLE

SILO

FEED ALLEY

PLAN

FEED CARRIER TRACK
← MANURE CARRIER TRACK

CROSS SECTION

F.H. KING (1900)

20.2 CIRCULAR BARN

as that desirable in human habitations, we suggest that you construct a south-facing, heat-absorbing masonry storage wall. Circulating air is heated as it flows between the heat-storing masonry wall and a double-layered fiber-glass collector panel. A sensor-controlled fan distributes heated air throughout the barn on cold but sunny winter days. At night, stored heat radiates inward from the masonry wall.

Hay and bedding material stored in the barn loft insulate the ceiling of the area below, which the animals inhabit. A bacterial litter colony provides floor insulation for this area. Deep litter is usually started in the early fall when there is still enough atmospheric warmth to ensure bacterial growth. Place a 6-inch layer of porous material, such as wood shavings or chopped straw, on the barn floor, inoculate it with a thin layer of old manure, and dampen it. As bacteria multiply and form a surface crust, add another 6 inches of dry bedding material, such as straw or leaves. As this material becomes soiled, add more litter. In the process of nighttime bedding and daytime use, animals compact this layer, expelling its air. This results in a slightly acid condition and slow fermentation, preventing the overheating, leaching, or drying out of the litter. Consequently, the nutritive value of the litter is maintained, so you can eventually use it as organic fertilizer-mulch in the garden, orchard, or field, while the interior barn atmosphere remains low in moisture content. It is this bacterial action and not litter absorption that keeps the moisture content of both the upper layers of litter and the barn atmosphere low. Poultry droppings, on the contrary, contribute large, unwanted amounts of moisture, necessitating additional ventilation, to be discussed shortly. Properly maintained, deep litter will not infect animal hooves and only has to be removed several times yearly.

Understandably, you might expect that earthen flooring would be more desirable under litter. Experiments at the Ohio Agricultural Experiment Station, however, disclose a 25 percent loss of liquefied excrement when earth floors are underlaid. Therefore, some type of moisture barrier is necessary under deep litter. You can mix equal parts of ground (hammer-milled) cardboard or sawdust and clay and equal parts of Bitumel (emulsified asphalt) and water in a cement mixer and trowel it directly onto the earth to make excellent flooring. If you use concrete, an insulating layer of air between earth and slab is desirable. You can create this air space by rototilling the earth to a depth of 8 inches. Then, after making holes with a crowbar, pour a series of vertical piles into the earth, creating concrete pillars on a 3-foot grid. Next pour a thin, 2-inch layer of concrete reinforced with wire netting over the tilled soil. In a few weeks the loosened earth will settle and the floor will remain suspended on its piles, leaving a continuous, insulating air space beneath. Called plunger pile flooring, this no-draft circulation is shown in Figure 20.3.

20.3 BARN SECTION
NORTH TO SOUTH

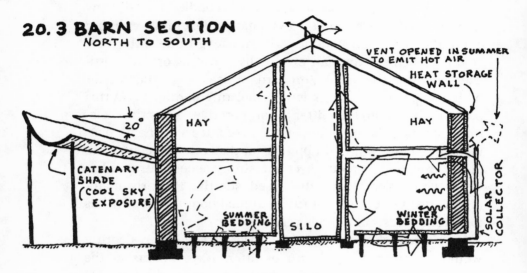

An adequate supply of fresh air is essential to healthy livestock. In winter this is as vital to animal metabolism as a supply of unheated air is essential to the proper combustion of a fireplace. But when exposed to cold, drafty conditions, animals, like fireplaces, tend to compensate their heat losses by increasing their food (fuel) consumption. The opposite result is apparent in summer months, when high temperatures may substantially reduce milk, meat, and egg production.

In his book *Bioenergetics and Growth,* Samuel Brody states that cattle dissipate as heat one-third of the energy of their grass feed. As atmospheric temperature rises, there is a corresponding increase in an animal's body temperature, resulting in less efficient feed utilization and less efficient breeding control. The problem of animal discomfort in hot weather is one of heat transfer. Four methods for cooling animals by heat transfer are conduction, convection, radiation, and evaporation. A properly designed barn and barnyard will provide for animal cooling in summer. Pigs, for instance, must rely heavily on evaporation because huge layers of body fat obstruct transmission of body heat to the atmosphere. Consequently, a pig requires a mud puddle or an occasional sprinkling to facilitate the evaporative process.

A year-round draft-free ventilation system is therefore critical to good barn construction. In winter louver-controlled air inlets (at the top of the opening doors) circulate outside air across the ceiling, where it is warmed before filtering draft-free throughout the animal quarters. Summer cooling is maintained by cross-ventilating through the large, open door-ways. A revolving-hood exhaust ventilator centrally located at the peak of the barn roof removes foul air by drawing it from the lowest regions of the barn, through the air space between the double-formed walls of the silo, and upward to the outside atmosphere. The suction power of wind exerts force on the fins of the hood, emptying the enclosure of used air.

The removal of excess moisture from the barn is also critical to the health of livestock. Excess moisture is best removed by natural ventilation (air flow). A temperature difference of 20 degrees Fahrenheit on opposite sides of a wall is necessary to create a flow of air through ventilating ducts. The King ventilating system, as it became known, consisted of flues in exterior walls for removing impure air and admitting fresh air. Air that has been breathed contains a high percentage of carbon dioxide and is consequently heavier than pure air. King placed used-air outlets near the floor of his barn, where impure air accumulated (Figure 20.4). As

TOP OF VENTILATOR TURNS ON STEEL PIN TO KEEP OPENING (C) ALWAYS TOWARD WIND. TAPERING CONSTRUCTION OF INTENSIFIER "SPEEDS UP" WIND AS IT PASSES THRU SO THAT IT LEAVES (A) AT HIGH VELOCITY. THIS CREATES SUCTION AT (F-F), EXERTING A POWERFUL PULL ON THE AIR IN THE OUTTAKE FLUE. LITTLE HINDRANCE IS OFFERED TO OUTFLOW OF AIR. AIR CURRENT HAS TO MAKE ONLY ONE TURN OF LARGE RADIUS (G-G, H-H). WHEN UPWARD SLANT OF WIND STRIKES (D) IT MUST BLOW DIRECTLY THROUGH AND OUT.

20.4 BARN VENTILATION

JAMES WAY VENTILATOR
1917

KING SYSTEM OF NATURAL DRAFT

A. FRESH-AIR INLET
B. CEILING VENT
C. FOUL-AIR OUTLET

WIND BLOWING IN
INTENSIFIER
HIGH VELOCITY
STORM BAND
SKIRT
OUTTAKE

foul air was naturally forced out, fresh air was drawn in through flue openings near the ceiling, where it was pre-warmed before reaching the animals' breathing level. King thus avoided cold drafts and promoted animal well-being. When functioning properly, King's system established a heat reservoir of warm air near the ceiling of the animal stalls, as illustrated in Figure 20.4.

In the 1950s, at the Swedish Institute of Technology, Gunnar Pleijel found that a clear sky results in an area of cool temperature at a compass point opposite the sun and approximately perpendicular to the sun's angle. Pleijel called this cooler area a "sink." On mornings the cold spot is in the west. Moving on a path counter to the sun's, it is in the east by evening. The mirrorlike reflection of the cool north sky explains why animals stand in the north shade of high-walled buildings, in preference even to trees. They are exposing their bodies to the radiant cooling effect of this so-called cold spot. Homesteaders can construct sunshades to take advantage of this effect. Properly designed, a sunshade will provide eastern and western sky shade, as well as noontime, northern sky shade.

Much research has been done on cold-spot cooling by Dr. Loren Neubauer of the University of California. At noon in California's Imperial Valley, the cold spot registered temperatures 28 degrees cooler than surrounding air. The north sky (noontime) cool spot was determined to be 25 degrees above the horizon, while the west and east (morning and evening) cool spots were about 40 degrees above this point. Neubauer found that a shade structure with a catenary form provided the most effective heat relief for animals. The structure sloped upward 20 degrees to the north and sagged between east-west supports, forming a curve (see Figure 20.3). This shade made the air as much as 20 degrees cooler.

In dry, western states a catenary shade should be about 12 feet high and about 8 feet high in humid southeastern states, where high cloud cover is found. Painting the top of the shade white and the underside black increases cooling. An installation for night-sky cooling is included on our homestead residence (see Figure 14.6).

The hypothetical barn and animal yard suggested in Figure 20.11 will likely take varying forms in different parts of the country, because topography and weather conditions greatly influence design layout. Specific climatic conditions also affect the insulation and ventilation requirements of individual barns, as do the availability of local materials and labor resources. Finally, the individual homesteader's preference

20. 5 EFFECT OF HEAT
AVERAGE DAILY GAIN OF PIGS KEPT AT VARIOUS TEMPERATURES

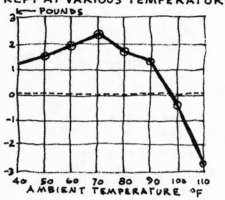

DECLINE IN MILK YIELD DURING SUMMER HEAT & DETERIORATION OF FOOD SUPPLY

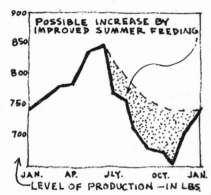

for certain numbers and varieties of livestock influences barn size and layout. We intend here merely to have you think about animal needs and basic animal-plant relationships.

A functional barn layout includes planning for inner yards, pastures, fields, feed storage, and waste disposal as well as the barn itself. King, for instance, was influenced by the arrangement of the traditional Norwegian barn, in which feed is delivered at the highest level, stored below the delivery level, and fed to animals at yet a third, lower level. Gravity disposal of manure takes place through the basement level. The basement composting level produces heat, which rises in cold weather to warm the animals, who are housed directly above. Overhead, thickly compacted hay in the loft insulates animals against the rigors of Norwegian winters.

King was possibly the first agriculturist to view the barn as only one of several elements in an animal's total physical environment, one that includes feeding, climatological factors, and management. The American Association for the Advancement of Science (publication no. 86) depicts the elements of an animal's physical environment as spokes in a

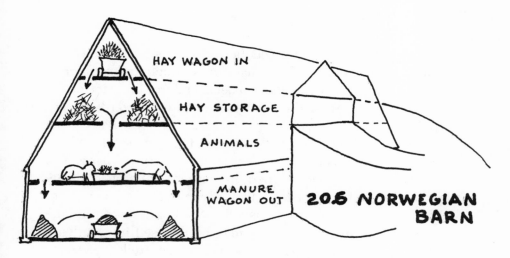

20.6 NORWEGIAN BARN

HAY WAGON IN

HAY STORAGE

ANIMALS

MANURE WAGON OUT

wheel. Man is represented as the axle, animals as the hub, and management practice as the lubricant that keeps the wheel in motion. The outside tread of the wheel is the total environment and is bound in place by a series of spokes representing elements of the physical environment that influence the well-being of the total complex. The delicate balance between the animal and its environment can be upset when any one of these spokes fails—that is, when any one of the physical influences creates stress. Improperly managed, a poorly lubricated wheel will come to a grinding halt. Abnormal air temperature is one factor that might distort the wheel, as illustrated in Figure 20.7.

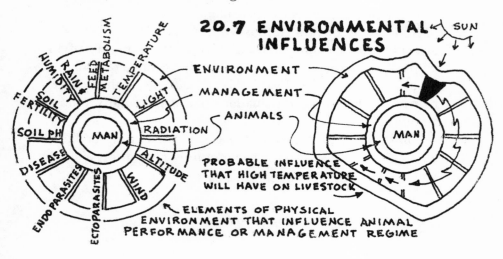

20.7 ENVIRONMENTAL INFLUENCES

SUN

ENVIRONMENT

MANAGEMENT

ANIMALS

MAN

RADIATION

METABOLISM
TEMPERATURE
FEED
LIGHT
RAIN
HUMIDITY
SOIL FERTILITY
SOIL PH
ALTITUDE
DISEASE
ENDOPARASITES
ECTOPARASITES
WIND

MAN

PROBABLE INFLUENCE THAT HIGH TEMPERATURE WILL HAVE ON LIVESTOCK

ELEMENTS OF PHYSICAL ENVIRONMENT THAT INFLUENCE ANIMAL PERFORMANCE OR MANAGEMENT REGIME

The model barn proposed here operates with three separate, alternately used animal yards. Each independently fenced yard has a pond, garden beds, and tree and vine crops. Animal access from the barn to any of these yards is controlled by opening or closing one or another of the barn doors. As each fish pond is yearly harvested, breeding stock and fingerlings are transferred to an adjacent gravity-fed pond. The vacated pond is then drained and either planted to a silage-producing sod crop or left fallow for animal use as an inner holding yard. Duck flocks usually accompany fish from one pond to another but at times may be released into outlying pastures. Goats and pigs are also allowed controlled access to pasture grazing areas. Vegetable crops are produced in raised beds in yards vacated by the animals.

A silo is the hub of our proposed barn-and-yard complex. Animal shelter surrounds the hub, which in turn is surrounded by animal yards and outlying pastures—in effect a wheel that rotates to produce a substantial yield of nutrient foods. It takes three years to generate a full rotation of this wheel in the following manner: The fallowed animal yard is planted to leguminous peas and beans, which enrich the soil with nitrogen and prepare it for root crops, carrots, turnips, beets, and the like. After the root crops are harvested, potatoes are planted in the naturally enriched, lime-conditioned soil. Finally, highly extractive plants, such as lettuce, cabbage, and cucumbers, are grown in soil that has reached its optimum fertility through cultivation of preceding nutrifying crops. After the harvest of greens, animals are again allowed access to this section of yard, while the former animal yard is heavily mulched in anticipation of the next early crop of legumes.

Similar crop rotation takes place in the two other fields. Every third year a grass-legume crop is grown for silage in

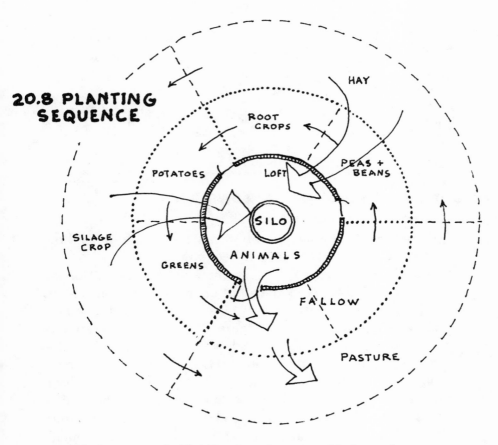

20.8 PLANTING SEQUENCE

HAY

ROOT CROPS

POTATOES

LOFT

PEAS + BEANS

SILAGE CROP

SILO

ANIMALS

GREENS

FALLOW

PASTURE

one of these two fields. To the silage is added surplus root crops, greens, and windfall fruit from the inner garden area. The other 1-acre field is planted, in rotation, with either fodder or hay, or it is used as pasture.

During fall and winter, when pastures are producing minimally, hay is fed as a supplement. To preserve these pastures from being overgrazed during winter months, silage feeding may begin as pastures are depleted. Figure 20.9 illustrates the relationship between the growth of grass and a

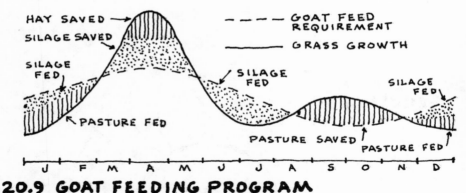

HAY SAVED →
SILAGE SAVED

SILAGE FED

PASTURE FED

SILAGE FED

PASTURE SAVED

PASTURE FED

SILAGE FED

- - - - GOAT FEED REQUIREMENT

———— GRASS GROWTH

J F M A M J J A S O N D

20.9 GOAT FEEDING PROGRAM

goat's feed requirement and how hay and silage may fill the gaps left by periods of minimal feed growth.

As Udall conjectures, "Bigger may not be better," for the enterprise that we propose can be organized on about 5 acres of land. Less may well mean more for either of the simple adobe structures shown in Figure 20.3 or Figure 20.10 may be used as the homestead residence while the family builds its income for future development. Living, cooking, and entry may be arranged below, while sleeping may be located above in what will eventually be the hayloft. The silo may be used as a two-level composting privy and bathing sauna. A family may be comfortably housed in this structure for as many years as it takes to develop the garden area, to build the shop and greenhouse, and to construct the final homestead residence.

The double-wall slip form used to build most other structures on this homestead may be used to build the innovative barn-silo using two thin, cast-concrete walls, which form a 1-foot-thick cavity between the two shells. Before each additional 4-inch layer is troweled into the form, the cavity space, except for ventilating air passages, is poured with an insulating mix of adobe clay. This type of construction holds winter heat and maintains summer coolness.

Another design of earth material is illustrated in Figure 20.10. Here, earth blocks made with a Cinva ram machine are laid against a revolving forming shoe. (Further details of this construction method may be found on page 113 of *The Owner-Built Home*.)

Figure 20.11 illustrates our own proposed model homestead farmyard for 3 acres of land. With two additional outlying fields of no less than 1 acre each, an average family living

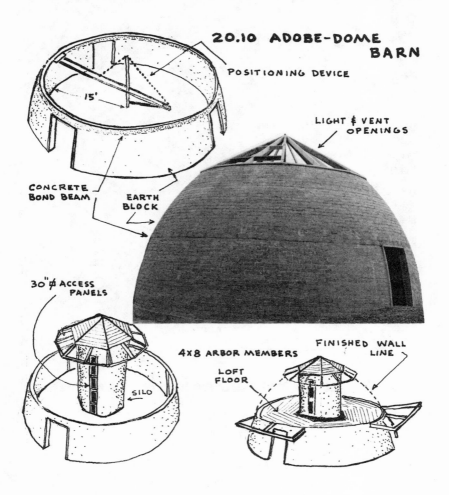

20.10 ADOBE-DOME BARN

POSITIONING DEVICE

15'

LIGHT & VENT OPENINGS

CONCRETE BOND BEAM

EARTH BLOCK

30"ø ACCESS PANELS

SILO

4×8 ARBOR MEMBERS

LOFT FLOOR

FINISHED WALL LINE

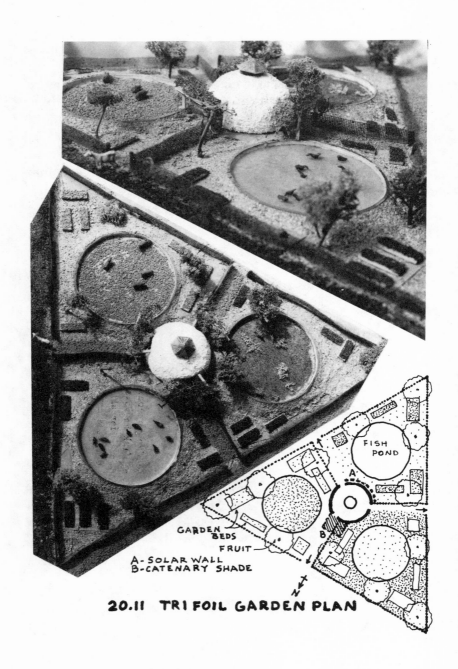

GARDEN
BEDS
FRUIT

FISH
POND

A - SOLAR WALL
B - CATENARY SHADE

A

B

N

20.11 TRI FOIL GARDEN PLAN

in a temperate climate should be able to produce a substantial portion of its total food requirement. Through intensive planting and careful management, this plot will support goats, ducks, pigs, and fish and will produce a year's supply of fruit, vegetables, and grain. Two outlying 1-acre fields are used alternately for pasture, silage, mulch, grain, and hay.

This proposal for a plant-animal-human ecosystem prompts the following question: How is the bacteria-laden, manure-saturated litter and homestead waste recycled back into the system? This question is considered in the following two chapters.

21

Fish Culture

Who is there who can make muddy water clear? If you leave it alone, it will become clear of itself.　　　　　　　　　　　　　　　　*Lao Tze*

　　　　Fish farming is an operation that is not only "clear of itself" but is one of long standing as well. The ancients of Mesopotamia, Egypt, China, and Peru fish farmed in their gardens, while in Europe this activity dates back to stone-lined Roman ponds. During the Dark Ages monks perpetuated this practice, supplying fish for sacramental observances. As a food-producing activity fish farming can be a simple, organic, low-energy operation. Therefore, leave it alone: do not complicate it and it will take care of itself.

Contemporary backyard fish farmers seem to feel that fish production requires a dome-covered pond containing exotic varieties from tropical sources that are fed and maintained by an elaborate feeding, heating, oxygen-supplying, and water-circulating system. Not so. In this chapter we describe how simply this program can be implemented on a homestead of any size. In Asia nearly every small homestead has a series of open ponds in which intensive fish production is conducted.

Fish farming is one of the truly exciting homestead occupations that is virtually self-managing. Generally unseen below the surface, wondrous breeding, feeding, growth, and decay activities occur. Inasmuch as this kind of farming produces valuable foodstuff, it is amazing how few North American homesteaders include fish management as part of their homestead program. Perhaps cultivation of the seldom-seen fish does not confer the same status and prestige that herding livestock bestows on the land farmer, but fish can produce twenty times as much protein as the land animal. On fertile permanent pasture young cattle may gain 300 pounds per acre per year; on infertile land the gain is about one-tenth as much. But fish production under intensive management practice yields, on the average, in excess of 6,000 pounds per acre per year. In terms of the increase in body weight for the weight of food consumed, fish gain twice the weight of cattle.

Fish protein is superior in quality to livestock protein; it contains no carbohydrates and only 0.1 percent fat. Being lower on the food chain, fish consume less energy in their life processes. A large part of the energy from food consumed by land animals is required just to maintain their body temperature and support their weight. Being cold-blooded and supported by water, fish are vastly more efficient energy converters and meat (protein) producers.

The need for status symbols only partially accounts for farmers' reluctance to consider fish production; there are other reasons besides ignorance of the value of this crop. Influential government agencies dominate this nation's agricultural programs. For example, the farm pond program started in the early 1930s by Franklin Roosevelt's administration offered a federal subsidy for farmers to construct conservation ponds. Even today, the Soil Conservation Service will pay as much as half the construction costs of a homesteader's pond. And when the pond is completed, it can be stocked with free fish from state fish hatcheries. But this type of pond construction does little to maximize fish production, and the handout of free fish is worthless because the varieties of fish the state hatcheries offer are often unsuitable for intensive fish culture.

The American conservation pond, therefore, must be differentiated from the Asiatic-European fish-culture pond, which is discussed later in this chapter. The conservation pond is best used as a water supply for culture ponds. Nevertheless, if well managed, it can yield as much as 500 pounds of fish per acre per year. From a carelessly maintained pond a farmer is lucky to land 100 pounds of fish a year. This is not fish farming.

Efficient conservation begins with a properly engineered pond. The watershed is the first resource to be considered when you entertain the idea of pond-building. In the eastern United States, where well-distributed rainfall averages 48 inches per year, the watershed can be as little as twenty times the surface acreage of the pond. In the West a larger watershed is necessary because of dry summers and high evaporative rates. However, you should avoid excessive inflow and flooding of the pond by building diversion terraces. There is really no foolproof way of screening fish from the overflow of a pond; a wide spillway tends, though, to save more fish than a deep one.

Your climate will determine the pond depth. A 3- to 12-foot depth is usually adequate in northern states, although when winter ice seals the pond, preventing an exchange of gases between air and water, you may avoid winter kill by deepening the pond to 16 feet. The added depth provides an additional supply of oxygen for the fish, which survive mainly because of their dormancy at this time of year.

Sufficient oxygen is supplied as much by the amount of pond surface exposed to the atmosphere as by any other factor of pond-building or management. In this instance a wider, shallower pond is preferable to a narrower, deeper one. Fish culturists find that fish stress is related to the oxygen content of pond water. Disease outbreaks occur after periods of low oxygen content in which a latent virus is activated.

When water temperature is the same from surface to pond bottom, as it is in early spring, equal water density promotes uniform water circulation and oxygen distribution. In summer, though, there is danger of what is called summer kill, caused by insufficient oxygen. The combination of cloudy skies, calm winds, and high temperatures in very shallow, weed-laden ponds can wipe out a fish population overnight. A shallow pond often becomes choked with aquatic vegetation, such as cattail and bulrush, if improperly managed. Quantities of plankton and algae also use up valuable oxygen.

A small but constant flow of water into a pond will raise its oxygen level. You should replace water from the bottom of the pond with fresh water, especially where temperature differences cause the stratification (or layering) of levels of the water. To ensure adequate circulation of upper and lower levels, install some sort of releasing device on the bottom of a conservation pond. Details of one such bottom-water overflow device are illustrated in Figure 21.1.

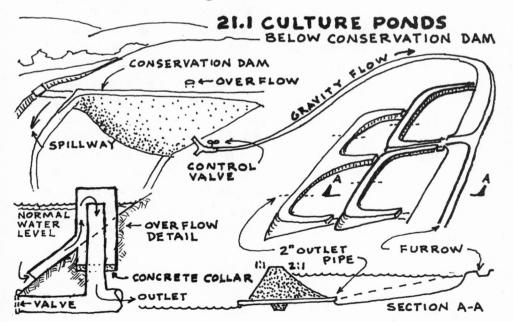

21.1 CULTURE PONDS
BELOW CONSERVATION DAM

CONSERVATION DAM
←OVERFLOW
GRAVITY FLOW
SPILLWAY
CONTROL VALVE
NORMAL WATER LEVEL
←OVERFLOW DETAIL
2" OUTLET PIPE
FURROW
CONCRETE COLLAR
VALVE
OUTLET
1:1 2:1
SECTION A-A

Asiatics and Europeans stock ponds with a single variety of one species of fish that utilizes all of the available food supply at various levels of the pond. Too little food may cause predacity among unrelated species of fish. Yet too much vegetation as well as too many fish, all requiring quantities of oxygen, causes any fish to excrete growth-inhibiting chemicals.

Some fish, such as grass carp, are stocked for the primary purpose of devouring grass and underwater weeds. It has also been learned that feces from grass carp contain large amounts of half-digested vegetable matter that other fish may feed upon. Fish manure, of course, will stimulate plankton growth. Grass carp do well in colder climates, and their eating of plants clears the shoreline, in effect increasing it. Progressive pond builders suggest building tree-planted islands, further increasing pond perimeters. Water insects, tadpoles, and crayfish live along shorelines and are welcome nourishment for the voracious appetites of such carnivorous fish as the trout, bluegill, catfish, and bass.

Stocking fish constitutes an important part of pond management, whether a conservation pond or a culture pond. Water temperature is the most important factor when spawning fish. Therefore, the species you stock should be chosen for its adaptability to your locale and pond environment. Largemouth bass, bluegill, and catfish thrive in warm water. Trout, however, require cold water, no higher in temperature than 60 degrees Fahrenheit, while yellow perch do well in water of intermediate warmth. Trout in general live relatively high on the food chain, requiring large amounts of protein for their development. In the West above 5,000 feet elevation and in the Pacific coast fog belt, ponds may be cold enough to raise trout. In the northern tier of states, the Appalachians, and New England, trout may also thrive, as illustrated in Figure 21.2.

The carnivorous bass lives relatively high on the food chain, too, as does the bluegill. But stocking practices have been formulated that make these fish wise choices for fish farming. They are stocked for mutual benefit at a ratio of 1 bass to 10 bluegill. Bass feed on bluegill during summer and fall and thereby control the bluegill, which otherwise reproduce extravagantly, threatening their own survival. In winter lowered water temperatures deactivate the digestive processes of bass, which then feel no hunger. Then, in springtime, bass, like all fish, are caught for human consumption at a time when they are most hungry—when the food supply is scarce. Summer is the time of year when bass are least available, for they have fed well on the prevalent bluegill, which itself feeds voraciously on summer-fall population of insects. A well-managed pond will yield 5 pounds of bluegill for each pound of bass.

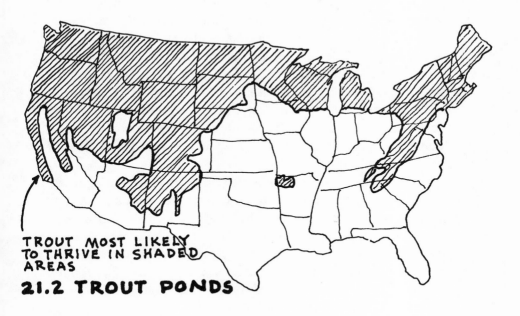

TROUT MOST LIKELY TO THRIVE IN SHADED AREAS

21.2 TROUT PONDS

Much work has been done to improve strains of fish for pond culture. One of the most promising of these improved strains is the *Tilapia,* a native of the warm African rivers. These fish are easy to breed, for a pair raised in favorable conditions will increase their number to 1.5 million in one year. Besides being largely free of parasites in crowded conditions, *Tilapia* live low on the food chain, have a herbivorous diet, and efficiently convert waste materials to foodstuffs. They do, however, require a warm environment in which to grow.

Fertilizing ponds can triple their yield. Fertilizer obscures their depths yet perfectly clear water denotes poor fertility. If you can see a white object a foot beneath the surface, a pond is not adequately fertilized. Organic fertilizer provides food for abundant growth of microscopic algae. Worms and larvae live on this algae, and in turn certain fish live on the worms and larvae. Indirectly, since fertilization of microscopic plants creates shade, nuisance-weed growth is minimized. Water weeds and shallow water plants should be discouraged, for they deplete the oxygen supply and cause stagnation.

In California hybrid Bermuda grasses are available from the Soil Conservation Service for the prevention of soil erosion along pond banks. These grasses will not clog the middle of ponds, although one variety will extend itself about 10 feet from the bank. This helps to break up the action of waves before they hit the bank in areas where wind occasionally whips pond surfaces.

Culture ponds differ from conservation ponds in four important respects: they are shallow, are smaller in size, always exist in conjunction with other culture ponds, and can be drained and dried up for a time. An ideal arrangement consists of three culture ponds, each ⅛ acre in size, terraced one above the other to receive drainage water by gravity. When stocked, these ponds should receive no through-flow of water, which would reduce the food supply. Slow seepage into the pond is preferred, mainly to keep the series full. Increase the 3-foot depth to 5 feet at one end where water is in limited supply or where winter frosts endanger shallows. In climates where ice will choke a pond, harvest the fish and drain the pond for the season, maintaining the breeding stock, however, in a smaller, separate, insulated tank. Monitor it for a balanced environment to ensure the survival of the stock.

A life of 350 million years on this planet has enabled fish to develop the capacity to outlive their predators. Even so, in the natural environment fish mortality is something like 80 percent. In culture ponds this may be reduced to 10 percent by maintaining the simple requirements for pond culture. These requirements for viable aquatic life are adequate space, stable temperature, enough food, enough oxygen, and a controlled population density.

To understand the dynamics of aquatic interrelationships, observe those of a fish, a snail, and a water plant in a bowl of water. The water plant uses snail waste produced by the snail's eating fish waste. And with a little help from the sun, the plant releases oxygen to purify the water and supply the snail and fish with this necessary element. The fish, in turn, lives off the plant. Unseen bacteria and algae also live at the end of this chain of events. The bacteria use complex waste materials in the water, while further along the food chain algae utilize inorganic salts, carbon dioxide, and water in the presence of filtered sunlight to produce simple proteins and carbohydrates. One form of plankton (phytoplankton) feeds on these basic nutrients. Another plankton (zooplankton) eats the phytoplankton, and minnows eventually eat the zooplankton.

Any homesteader can duplicate this kind of balanced environment on an economic scale. Farmers in the Rhine Valley have been doing so for several hundred years. After growing grain for a few seasons, they flood their diked fields and stock them with carp. The carp thrive on vegetative residue, mature rapidly, and are harvested when the field is eventually drained. The fish-manured fields are then resown to grain, and the cycle is repeated.

Modern homesteaders could do no better than pattern their fish culture after that of the venerable Chinese system. It works and has for some time. Several thousand years ago Chinese farmers perfected pond culture, developing uncanny techniques for water control and dam construction as well. They knew almost everything known today about breeding and stocking fish, pond fertilization, and weed control. Their terraced ponds had many advantages, not the least of which was the separation of the breeding and growing stages of fish development. One of their greatest contributions was the breeding of varieties of one species of fish to live on the various foodstuffs found on the different levels of a balanced aquatic environment. They stocked herbivorous, plankton-eating, and bottom-feeding varieties. They even bred carp to produce varieties of this species that would consume a specific diet at a specific level of the pond.

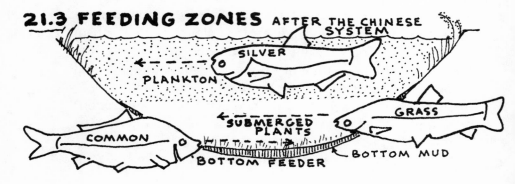

21.3 FEEDING ZONES AFTER THE CHINESE SYSTEM

SILVER

PLANKTON

GRASS

SUBMERGED PLANTS

COMMON

BOTTOM FEEDER

BOTTOM MUD

21.4 POLYCULTURE

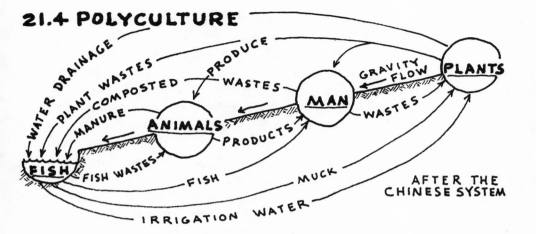

AFTER THE CHINESE SYSTEM

Under present-day Chinese management, fish culture is an integral activity that includes in its production cycle the orchard, garden, pig, duck, and chicken as well. This program of polyculture begins with the terraced ponds, where water is impounded by gravity and where nutrients are drained from garden and orchard above. Pig and poultry manure is also washed into the first of these ponds. Pig dung contains 70 percent digestible food for fish. When the fish are harvested, the offal and the undersized are fed in turn to pig and poultry.

Most Chinese ponds are drained every few years. Aquatic vegetation and surplus muck are scraped up to fertilize gardens and orchards, and pond bottoms are exposed for a period to sun and air. Pond bottoms are then planted to soybeans, rice, or alfalfa. After harvest the pond is refilled and restocked. This practice keeps fish disease and parasitic incidence at a minimum.

When filled and restocked, nutrient-laden overflow enriched with fish manure is used to water crops. In China only the immediate soil around the plant is fertilized, not the entire field. The comprehensiveness of the Chinese system is glimpsed in their custom of providing privies over their ponds, inviting passersby to stop to relieve themselves.

Ducks enjoy a symbiotic relationship with fish, for they feed on invertebrates and on vegetation fertilized by the fish. The Muscovy breed seems to be the hardiest and the easiest to raise. It has been found that carp raised in ponds to which ducks have access grow from two to five times faster than those lacking association with ducks. Ducks, along with frogs, chickens, and pigs, perform a kind of natural pest control, thriving on this source of nutrition. There is thus an apt corollary to the old saying "If you can't beat 'em, join 'em," to the effect that "if you can't eradicate pests, use 'em."

In recent years FAO has done much research on the intensive cultivation of fish and cold-blooded aquatic invertebrates, such as mussels, clams, and crayfish, in optimum growing conditions. This science, called aquaculture, is an important aspect of homestead polyculture, for it facilitates the management of a variety of crops in a single production area. In 1966 the United Nations held a world symposium on warm-water fish-pond culture, and the five-volume proceedings was subsequently published. This is a useful reference for the pond-operating homesteader and is available in most large libraries.

Pond culture is developing under guidance from an impressively qualified scientific community. One privately funded group, the New Alchemists of Woods Hole, Massachusetts, has been involved in experimental work to develop backyard fish farming, and as homestead aquaculture gains in popularity, these people will be well qualified to provide essential coordination of the various aspects of this important work.

Currently, in the United States, commercial production of fish from culture ponds or tanks is developing much as have other monoculture-farming and livestock-raising practices. Commercial fish raisers seek maximum profits from maximum yields with devices amounting to fish feedlots. In these, hapless creatures are fed high-protein rations manufactured by major livestock-feed companies. Yet for all of these efforts, methods employed by simple farmers of the Orient consistently outproduce those used in the West. Mainland China today produces more fish from ponds than any other nation in the world—over 1.5 million tons in 1965.

The commercial devices mechanically operate all aspects of the aquatic environment; they are closed systems. To keep the packed fish from dying in their waste, the forty-tray fish-producing unit pictured in Figure 21.5 requires conditions in which water temperature and oxygen levels are kept constant at all times. This production unit acts as a "heat sink" and probably uses up more thermal energy than the food energy it produces. The harvest from these trays amounts to an

CONTROLLED ENVIRONMENT   CULTURE POND
21.5 AQUACULTURE

equivalent of that from ponds containing 30 acres of water surface. Operators claim that if all goes well, this single installation can produce an estimated 40,000 tons of catfish yearly.

You may anticipate the extent to which such measures will go when these practices include controlled marketing features as well. When fish reach 1¼ pounds, if there is no need to market them immediately or if prices are unfavorable, it is possible to reduce the water temperature and withhold feed. Fish will hold their weight, marketers boast, "until market conditions are more favorable." Already modern artificial fish propagation involves experimentation with induced breeding and hormonally induced spawning using intramuscular injections of pituitary extract.

In California and Louisiana, another example of commercial fish production takes place in open systems of ponds stocked with ⅛-inch catfish fingerlings. The fish are either raised in static (not periodically changed) water or in pond water managed for regular change; the first method delivers 1,500 pounds per harvest per surface acre and the second yields 2,700 pounds of fish per surface acre. Harvest weights are usually about 1 to 2 pounds each. Depending on when ponds are stocked, these fish may be harvested from April through June and from September through November.

22
Sanitation

We must base our next industrial revolution on using our wastes. Wastes are simply useful substances that we do not yet have the wit to use.
Athelstan Spilhaus

The final chapter of books on country living often begins the discussion of sewage disposal. We prefer to speak of usage rather than disposal, for sewage is not a commodity to be casually flushed away as something distasteful and valueless. It is entirely possible and highly desirable that all homestead wastes be managed like any other products of the fully functioning, recycling homestead. Such management produces usable fuel, animal food, and soil enrichment once the decomposition of pathogenic bacteria and parasites in waste has destroyed them or rendered them harmless.

In any process of organic-waste decomposition, nitrogen is the primary chemical agent. Nitrogen in its gaseous state is abundant in air. It is an inert (inactive) element but is nonetheless essential for tissue growth and repair in higher plants and animals. Either the voracious appetite of uncounted numbers of microbes or the catalytic energy of lightning are required to fix (or combine) nitrogen with other elements, making it available to plants for their protein manufacture. This process is called nitrification. Plant protein eaten by animals is converted to animal protein, forming nitrogen-containing acids, the amino acids, which are used in tissue-building. Then, through biological processes, plant and animal tissues decay and nitrogen is freed from its bond with other elements and is restored to the atmosphere. This is denitrification.

Microbial activity in this cycle is known as the biology of digestion. Microbes synthesize compounds of nitrogen, which eventually become living plant or animal tissue, or they digest them in the process of decay. Whereas nitrification immobilizes (captures) nitrogen for protein synthesis, in the process of decay denitrification releases nitrogen from its captivity for reuse. In this process pathogens and parasites are inactivated.

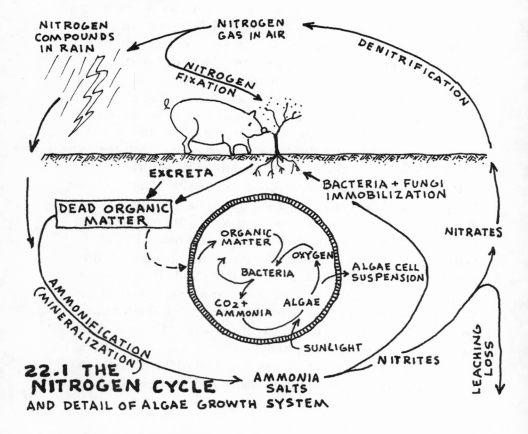

22.1 THE NITROGEN CYCLE AND DETAIL OF ALGAE GROWTH SYSTEM

Two kinds of microbes exist to decompose fixed compounds of nitrogen. Anaerobic bacteria function in the oxygen-deficient watery medium of all septic and sewage treatment systems. Since there can be little consequential heat buildup with this method, it takes up to six months to destroy harmful organisms in the waste material. Aerobic bacteria, on the other hand, can destroy pathogens in a matter of hours when waste is subjected to the high temperatures generated by the oxidation of organic material in a recycling composting system.

One method for decomposing waste employs anaerobic bacterial activity to produce a supply of methane fuel for the homestead, a high-protein animal food, and a stabilized sludge for soil enrichment. The methane-digester method is, however, more than a septic system with an open leach line. It is part of an ecologically sound food-to-waste-to-food program.

The senior lecturer in environmental health at the University of Papua, Port Moresby, New Guinea, George Chan, has developed a working demonstration of a waste-recycling program. A schematic diagram of this water-waste-fuel-food cycle is shown in Figure 22.2. Human waste (1) and animal waste (2) are washed daily into a digester tank (3), where anaerobic decomposition takes place, producing methane gas, which is piped to the house. Effluent from the digester and its settling tank (4) is discharged into a shallow pond (5), where sunlight and warmth cause high-protein algae to grow. Some of the algae are harvested and fed to pigs and poultry. The purified effluent is then channeled into a fish pond (6). Herbivorous warm-water fish feed on organic matter and are harvested for human consumption or for nutritious animal feed. The final, stabilized sludge from the digester and the settling tank is placed on the garden (7), where it enriches

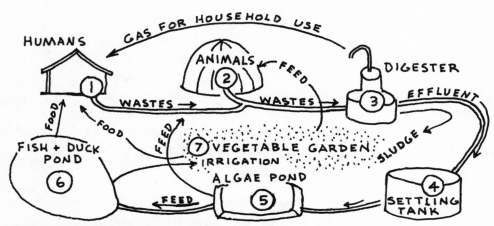

22.2 WATER-WASTE-FUEL-FOOD CYCLE

growing vegetables and trees. Mineral-rich pond water is used for irrigation and fertilization of garden and orchard. Models of this digester have been installed in both Fiji and New Guinea, and the Taiwan government has sponsored over five thousand of these units to promote rural development and prevent the drift of young people to cities.

Specifically, a digester is a simple construction of metal or concrete, a 300-gallon tank suitable to rural installation. Built below ground (or insulated if built above ground), it is covered with an airtight cover of $1/8$-inch sheet iron painted black and utilizing solar heat to promote digestion. Temperatures thus reach 160 degrees Fahrenheit, destroying all pathogens. This heavy covering pressurizes the gas that is produced by waste decomposition within the tank. Gases that are formed are 60 to 70 percent methane.

The tank may be supplied with a watered-down "waste slurry" at either regular, daily intervals or in one single loading. Fifty cubic feet would be the minimum digester volume for this system on a small homestead. A population of thirty ducks, three goats, three pigs, and five people creates about 25 cubic feet of waste material daily, a minimal amount for efficient operation. (When animals are given free range, only a portion of their manure can be recovered.) In Melanesia waste from this digester will deliver about 10 cubic meters of gas daily—sufficient for household cooking, refrigeration, and illumination.

The capacity of the settling tank must accommodate a 24-hour flow of effluent. Again, anaerobic bacteria digest organic matter and destroy persisting harmful organisms. Effluent and scum from these tanks then flow into a small holding pond of puddled clay or concrete with sides 1 foot high. This

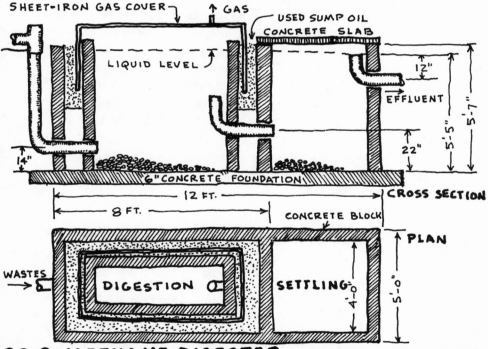

22.3 METHANE DIGESTER DESIGNED BY GEO. CHAN

shallow depth allows the maximum penetration of sunlight, which contains germicidal ultraviolet rays. These sanitizing rays eliminate any vestiges of pathogens. Algae, one type of single-cell protein, flourish in this medium and should be removed daily and mixed with animal feed or sun-dried for future use.

Single-cell protein is the result of a highly efficient method of food production. It takes a radish plant a month and some livestock several months to reach food-useful size. A single-cell alga doubles its weight in half a day. Yeast is even more efficient, doubling its weight in only 2 hours. By comparison a 1,000-pound bull can synthesize less than a pound of protein a day, whereas 1,000 pounds of yeast will produce 50 tons in the same amount of time. This makes the production of yeast more than a hundred thousand times as efficient as the raising of a bull.

Natives of the Republic of Chad have used algae as a staple of their diet since prehistoric times. Cortez, likewise, observed in 1521 that, near what is presently Mexico City, natives "sell some small cakes made from a sort of ooze which they get out of the great lake, which curdles, and from this they make a bread having a flavor something like cheese." The blue-green alga that Cortez referred to has been identified as *Spirulina,* the same variety that Chad natives to this day collect from oases for their cakes.

SOURCE OF PROTEIN	COMPOSITION			COMPARISON OF WATER USE		
	% PROTEIN	FAT	CAR.			
SOY BEAN	46.7	7.1	40.9	576	2.0	288
CORN	10.2	3.9	85.2	240	2.0	120
WHEAT	13.6	1.5	84.1	135	1.5	90
ALGAE	63.1	4.4	2.1	20,000	4	5,000
EGG	48.8	44.5	2.8	ANNUAL PROTEIN YIELD LB/ACRE	ANNUAL WATER CONSUMED AC.FT/AC	POUNDS PROTEIN PER ACRE FOOT
MILK	26.9	30.0	37.7			
FISH	55.4	37.9	—			
MEAT MUSCLE	57.1	37.1	2.0			

22.4 ALGAE FOOD VALUE

Purified effluent from the settling tank is drained into a fish pond about 50 feet in diameter and 5 feet deep. Fish feed on water plants, plankton, and algae. (There is no vestige of human or animal excreta left in the system.) The fish are in turn eaten by humans or processed for animal feed. Ducks eat water weeds, mosquito larvae, and invertebrates. Together with fish they keep the pond clean.

The garden and orchard bloom when watered and fertilized by excess water from the fish ponds. Humus is made from sun-dried sludge, which amounts to several buckets per person per year. Vegetables grown in soil enriched with this material provide part of the animal diet. Ultimately the pond is drained and its bottom scraped of the stabilized sludge that collects there. During alternate seasons the pond bottom is planted to a silage crop.

Professors Golueke and Oswald of the University of California have adapted this system for contemporary use, with a digester located in the center of a circular structure that houses both animals and people. In this system effluent overflows directly into a pond located on the roof (see Figure 22.5).

Producing algae on a rooftop, as done in the system used by the Golueke-Oswald design, is no more complex than the production of any crop. Sunlight, water, and supernatant are all that is required for the process. The water consumed for algae production may be annually twice that for other crops, but the protein yield from algae is considerably higher than that of other crops, as Figure 22.4 shows.

Perhaps it is now clear to the reader that recycling excreta is not only possible but desirable. Another method of decomposition exists that is also efficient and produces stabilized soil nutrient. This is the composting method of excreta usage. We proceed with this discussion of sewage treatment on the assumption that excreta is not waste at all.

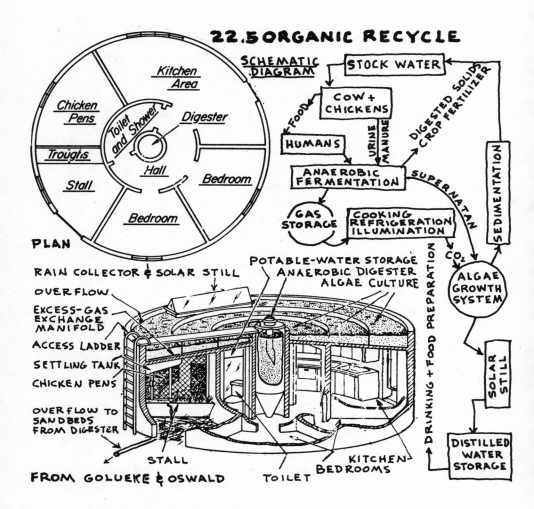

22.5 ORGANIC RECYCLE

SCHEMATIC DIAGRAM

STOCK WATER

COW + CHICKENS

FOOD

HUMANS

URINE MANURE

DIGESTED SOLIDS CROP FERTILIZER

ANAEROBIC FERMENTATION

SUPERNATAN

GAS STORAGE

COOKING REFRIGERATION ILLUMINATION

SEDIMENTATION

CO₂

ALGAE GROWTH SYSTEM

DRINKING + FOOD PREPARATION

SOLAR STILL

DISTILLED WATER STORAGE

PLAN

Kitchen Area

Chicken Pens

Toilet and Shower

Digester

Troughs

Stall

Hall

Bedroom

Bedroom

RAIN COLLECTOR & SOLAR STILL

OVERFLOW

EXCESS-GAS EXCHANGE MANIFOLD

ACCESS LADDER

SETTLING TANK

CHICKEN PENS

OVERFLOW TO SANDBEDS FROM DIGESTER

POTABLE-WATER STORAGE
ANAEROBIC DIGESTER
ALGAE CULTURE

STALL

TOILET

KITCHEN BEDROOMS

FROM GOLUEKE & OSWALD

The pit privy, in common usage rurally for some time, is simply a covered receptacle in the earth for the deposit of untreated organic matter. This may in some areas leach dangerous pathogens into groundwater supplies. Its successor, the septic system, is costly, wasteful, and polluting—but no more gross than our urban sewage systems. The alternative to these is simple in concept (thanks to the pioneering work of WHO), ecologically sane, and reasonably inexpensive. Modifications of its cost upward or downward depend entirely upon each homesteader's particular financial means. We present in Figure 22.6 a schematic representation of a composting privy recently built and in use at our new homestead. This unit is the result of a dozen years of progressively successful experimentation. It will suffice until time allows us to develop the full-cycle fuel-fertilizer-food system described above.

The fundamental feature of the privy's design is its basic function within the overall homestead system. It nestles between greenhouse, cooking, and utility areas and close to sleeping accommodations: The composting chamber is additionally loaded with organic matter from the garden and environs and is emptied from the chamber through an outside, lower-level access convenient to garden and greenhouse. Figure 22.6 illustrates this arrangement and includes in its design an upper-level sauna, a bathing tub, and water storage and can accommodate a tower for the wind generation of power. An inexpensive, single-level privy can, however, be built to serve for washing, bathing, and evacuation.

Aerobic decomposition takes place in a divided chamber that is a cast, double wall of concrete. Twice a year a simple metal baffle plate is turned to divert excreta material from one side of the chamber to the other, and twice a year alternate chambers are harvested of their stabilized humus in preparation for a new batch of composting organic material and excreta.

Putrefaction releases ammonia. Therefore, any composting of organic material requires management. Too much urine, for example, raises the nitrogen level, resulting in an odorous condition within the system. This can be controlled by adding

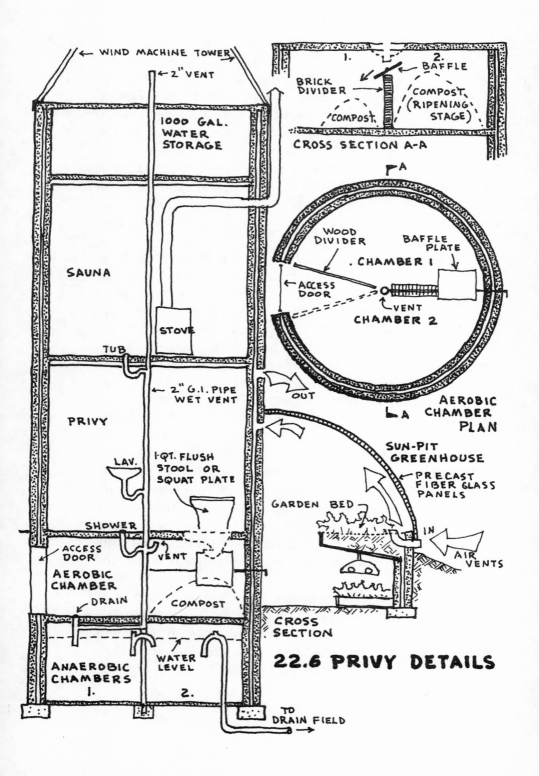

← WIND MACHINE TOWER

← 2" VENT

1000 GAL. WATER STORAGE

SAUNA

STOVE

TUB

← 2" G.I. PIPE WET VENT

PRIVY

LAV.

I-QT. FLUSH STOOL OR SQUAT PLATE

SHOWER

ACCESS DOOR

AEROBIC CHAMBER

VENT

DRAIN

COMPOST

ANAEROBIC CHAMBERS 1.

WATER LEVEL

2.

TO DRAIN FIELD

1.

BRICK DIVIDER

COMPOST

2.

BAFFLE

COMPOST (RIPENING STAGE)

CROSS SECTION A-A

A

WOOD DIVIDER

BAFFLE PLATE

CHAMBER 1

← ACCESS DOOR

VENT

CHAMBER 2

AEROBIC CHAMBER PLAN

A

OUT

SUN-PIT GREENHOUSE

PRECAST FIBER GLASS PANELS

GARDEN BED

IN

AIR VENTS

CROSS SECTION

22.6 PRIVY DETAILS

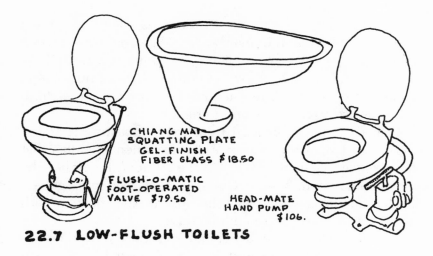

CHIANG MAI
SQUATTING PLATE
GEL-FINISH
FIBER GLASS $18.50

FLUSH-O-MATIC
FOOT-OPERATED
VALVE $79.50

HEAD-MATE
HAND PUMP
$106.

22.7 LOW-FLUSH TOILETS

carbon-containing materials—that is, cellulose materials such as paper, sawdust, or leaves. Humidity in the mass is also important, requiring proper management. Some moisture is necessary to provide aerobic organisms with a soluble food supply. Too much moisture, however, reduces air spaces in the working mass, limiting the oxygen used by these microbes for waste digestion. As a rule, moisture should be maintained at 50 percent. This percentage includes feces, which is 70 percent moisture; urine, which is 90 percent moisture; organic refuse, garbage, and vegetative trimmings, which are 20 percent moisture; and a flush of 1 or more quarts of water.

The composting privy system of waste recycling is an immensely practical consideration when compared to the standard 5-gallon water-flushed septic or sewage-disposal systems. This alternate convenience uses 4½ gallons less than our present water-wasteful devices. A recent booklet, *Stop the Five-Gallon Flush!*, by the Minimum-Cost Housing Group of McGill University's School of Architecture

(Montreal, Canada), describes numerous water-conserving methods for handling human waste. The various units listed show the chemical, recirculating, vacuum, freeze, and incinerating methods of waste disposal presently being developed or in use. Interested readers can write to us for a complete listing of water-conserving toilets.

All bath water is diverted from the composting chamber and drained into a separate anaerobic treatment chamber that destroys pathogens remaining in the effluent, much like the conventional septic system. Overflow is released into a gravel-lined drain field, which may be built to specification with county sanitation department's blessings.

Hot- and cold-water faucets and a lavatory fixture are attached to the privy's central 2-inch galvanized pipe, which handles drain water and doubles as a compost air vent. It is important to remember that plenty of air is critical to speedy aerobic decomposition. To destroy pathogens and parasites, a lot of air is needed to raise the temperature of the organic material to a sweltering 140 to 180 degrees Fahrenheit. Presently we are considering using a vibrating grid of blades strategically placed to receive and cut donations into pieces for better mixing with previously deposited aerating materials.

The central pipe of the system provides alignment during building construction. From it the versatile slip form is swung on its semirigid guide arm. Concrete is troweled into the form, creating curvilinear walls. When finished, this same pipe serves as the anchor for floor and roof slabs. The core between walls is insulated with styrofoam beads. In severely cold climates it is particularly important to insulate the compost chamber, which must remain quite warm at all times for decomposition to take place.

Concrete was chosen as the basic construction material. It is requisite for any sections of the structure to be built below ground. The sauna interior may be lined with fragrant, moisture-absorbent wood, such as cedar or redwood.

Homesteaders are well advised to consult with the local sanitation department about constructing a composting system for sewage treatment if they are building in code-enforced districts. However, many homesteaders are fortunately located outside such jurisdictions, where approval is not mandatory. Others, armed with authoritative literature from WHO and other qualified sources, such as *The Owner-Builder and the Code,* are doing legal battle with city hall to regain the right to utilize and conserve this resource as well as all of our natural resources in a safe and sane way.

Over the decades a number of healthy minds in the fields of medicine and agriculture have studied sewage treatment to try to find a more rational system of sanitation. One of the earliest of these was a modest English doctor named Poore who wrote in 1894 a book called *Essays on Rural Hygiene.* George V. Poore unabashedly raised vegetables on 1½ acres of land fertilized by fresh donations of human excrement. Wastes from a hundred persons were covered with soil, requiring four years to dress the entire garden area completely. Poore practiced this husbandry for twenty-two years, after which time he wrote, ". . . fertility and beauty of the garden have been increased enormously. No other crops, except cabbage, seem to flourish in fresh material, but cabbage may be followed by potatoes, these by celery (planted between the rows), these by peas or beans and, after this, by parsnips or carrots, without any fresh manuring and with a most abundant yield." Poore's spirited words in support of his actions are in the best tradition of the well-informed freeholder:

That in a country or semirural district, where it is possible to give a house a decent curtilage or small garden, it is easy for a householder to make the sanitation of this dwelling quite independent of the local authority. In fact, the householder is able, if so minded, to make his sanitation complete and to furnish, on his own premises and to his own profit, that "circulation of organic matter" which is the law of nature and the only true basis upon which the science of sanitation can possibly stand firm. The householder can do piecemeal what no public authority has ever succeeded in doing wholesale, albeit that millions of pounds have been wasted in silly attempts.

Although Japan, China, and India have for centuries composted animal and human waste in a similar manner, it is only since the 1930s that certain Western nations have considered treating organic wastes for their reuse. The Netherlands, Sweden, and Israel are now among the countries of the world that can no longer afford to destroy this storehouse of nutrients. Annually these countries are producing thousands of tons of fertilizer from municipal waste, and India alone produces more than 2 million tons of this indispensable wealth. Perhaps the world's people are beginning to be concerned about our profligate waste of irreplaceable resources, such as nitrogen, phosphorus, and potassium. It's high time . . . or perhaps it's later than we realize.

23
Food Preservation

The very idea of being fed, or a family being fed, by daily supplies has something in it perfectly tormenting. William Cobbett

For centuries farmers have been concerned with various methods of preserving some of their crops, indignant to think that they would be "fed by daily supplies" from sources other than their own handiwork. Perhaps the initial discovery that sun-dried cereal grains could be kept through winter months encouraged wider experimentation and application of food-preservation techniques. In any event, man certainly could not get on with the business of developing a civilization as long as much of each day was required just to locate food sources and chew tough, raw meat. Anthropologists, therefore, correlate advanced society with advanced methods of preserving food.

Unknown until recent times, the basic principle of preservation is to create an environment unfavorable to the growth of microorganisms, such as bacteria, molds, and yeasts, which cause noxious changes in the appearance and the chemistry of foods. Extremes of heat or cold, lack of water, excess saltiness or acidity, and the presence of chemical additives all discourage or destroy food-spoiling organisms. Some of these organisms are bacteria that produce toxins. Certain enzymes naturally present in all plants and animals may also cause spoilage. Therefore, one objective of preservation is to stop microorganic and enzymatic activity. Some enzymes, however, do have nutritional value. Eskimos, for instance, bury fish to activate an enzyme that produces the semiliquid tilnuck, a malodorous product unsuitable to non-Eskimo taste but considered delectable by these northerners. Heating alters enzyme chemistry, enhancing the taste and texture of foods eaten shortly thereafter.

A list of the ways in which people, over the centuries, have preserved food is interesting to peruse. First came sun-drying, fermentation, cooking, and smoking meats. Then sugar, salt, and fruit acids, such as acetic acid (vinegar) and citric acid, came into use. In more recent centuries we discovered pasteurization for bottling milk and fruit juices and canning for keeping a variety of foods. Dehydration, refrigeration, and freezing are contemporary means of conserving foodstuffs, along with the widespread use of chemicals.

This last method of food conservation would be "perfectly tormenting" to the likes of William Cobbett. He would have found reprehensible gasing dehydrated fruits with sulfur dioxide, destroying vitamin B_1 to retain C; inoculating poultry and fish about to be refrigerated with antibiotics; adding sodium benzoate to acid fruit juices and syrups and margarines; impregnating low-acid juices, such as apple and grape, with carbon dioxide; and using propionates to check the formation of molds on bread and other compounds that retard the oxidation of certain fats. Dates and figs are treated with propylene oxide to prevent their souring and fermenting. Water is partially removed from foods, and milk and juices are concentrated and heat-processed or refrigerated for their shelf life. The newest miracle method emanates from General Electric's X-Ray Department. Various foods are being bombarded with sterilizing doses of ionizing radiation to inactivate enzymes and destroy bacteria. It is not known what changes this process brings about in carbohydrates, protein, and fats, but such irradiation extends the storage life of many fresh fruits and meats and prevents potatoes from sprouting and bakery goods from molding.

Cooking, along with sun-drying, became an early means of preserving seeds. The product kept for a number of weeks merely by cooking and recooking it regularly. In relatively recent times it was found that the vitality of heavy, textured flat breads could be extended by adding yeast. As a result, grain pastes became leavened bread. It was probably by accident that cheese became the preserved form of milk. Stored in a pouch made of a calf's stomach, which contained the enzyme renin, milk reached semistability when it curdled.

Primitive herders were confronted with one major problem involving their traveling meat supply. The herd had to be kept to a number that could be fed between one harvest and the next. Otherwise, as feed supplies dwindled, large numbers had to be slaughtered and their meat somehow preserved. Smoking meat was one of the primary methods used. Smoked meat and fish, such as the kipper, resisted spoilage when surface tissues were simultaneously dehydrated by heated air and impregnated with preservative chemicals, the distillation products of slowly combusting wood.

Three thousand years ago the Egyptians preserved meat by rubbing it with salt. Bacteria that cause spoilage are kept under control by this method because salting withdraws water from cells. Vegetables cured in a salt brine are fermented by the formation of lactic acid. In this manner cabbage becomes sauerkraut. Only part of the salt penetrating meat tissues can be removed by boiling, making this a questionable practice. One laboratory study has discovered that ingesting 1 ounce of salt a day—not an uncommon dosage— can shorten a person's life by as much as thirty years!

Smoking and salting meats may have been necessary when

animals had to be slaughtered because of a lack of winter feed, but we question how justifiable these methods are today, when slaughter on the homestead can be regulated to need. Currently we have far more efficient methods for meat preservation, methods that render a more nutritious and less toxic product.

Protein is best cooked quickly because its amino acids and its B vitamins deteriorate when subjected to long periods of heating. Broiling and quick searing for roasting are better methods for cooking meat than boiling or frying. The muscle fiber of meat is held by connective tissue. When properly cooked, connective tissue softens and muscle protein becomes firm. When cooked with excessive heat, muscle protein rapidly shrinks and compresses. Connective tissue breaks down, softens, and dissolves, forming a gelatinous mass. It must have been known for a long time that any one of the over two dozen varieties of meat is tenderized by hanging it in a cool place, allowing time for enzymes in the flesh to ferment the protein slightly, softening the tissue. Pork, veal, and lamb, however, should be processed as soon as they are butchered.

Heating destroys essential vitamins and minerals in all foods. In pasteurization food is heated to near boiling long enough to destroy organisms. Canned food is sealed in containers and is subjected to heat. Nonacid foods, such as vegetables, meat, fish, poultry, and milk, require higher temperatures than acid foods with a pH lower than 4.5. Vegetables must be blanched or scalded before processing by heating, freezing, or dehydration to check enzyme activity. Vitamin-C losses can be as high as 75 percent during meal preparation, including cooking. Rapid boiling of vegetables in quantities of water utterly destroys vitamin C. Water-soluble vitamins, such as C and B-complex, cannot survive the twin injuries of exposure to heat and water. Vitamin-A deficiency is not so much the result of its destruction through rigorous cooking as it is a shortage of unhydrogenated fat in the diet.

There is one method of frying food used by the Chinese that we find acceptable. They use a cooking vessel, the wok, for a kind of intense, brief sautéing called stir-frying. A "waterless" cooker par excellence, the wok has tapered sides, which reflect heat toward the center of the container where rapid cooking under relatively high heat is accomplished in a mere tablespoon or two of fat. Vegetables are especially suited to wok cookery because they normally contain high percentages of water.

Time takes its toll of nutrients too. After three months of cool storage, potatoes lose half their ascorbic-acid content. Nine-tenths of what remains is lost as a result of cooking and reheating. Not much food value greets the consumer of the good old American restaurant hash-browns breakfast special! Fresh blueberries have 280 units of vitamin A, but when they are canned in heavy syrup, this amount is reduced to only 40 units. A ripe apple has 90 units of vitamin A, but as sugar-cooked applesauce it has only 20 units. These same fruits in their natural sour state lose little vitamin C, because their acid content inhibits the enzyme activity that leads to vitamin losses.

Sour fruits are made into jams, jellies, and preserves, but only after being heated in huge quantities of sugar, amounting to 70 percent of the total amount of preserves. The moisture content of these fruits is high, above 15 percent, a suitable environment for spoilage organisms. Sugar acts as a preservative by creating an acidity in which these organisms cannot grow. However, a valuable booklet by Dr. Michael Jacobson called *The Nutritional Scoreboard* gives all sugar products a minus rating.

Any carbohydrate (starch or sugar) can be fermented, a natural means of preserving foods. The quantities of sugar used to preserve foods can be reduced by decreasing the amount used and by adding yeast. Yeast enzymes convert sugar to alcohol, changing sugar-laden fruit juice into wine. To make beer, starch must first be converted to sugar by enzyme action or hydrolysis. In this latter process grain is allowed to steep and germinate; then it is dried and ground into malt. The process that usually renders a food spoiled can thus be controlled to produce a stabilized beverage. Wherever a stable society has existed, there is evidence of fermented beverages and foods.

Dairy products spoil as a result of microbial activity, yet cheese is a result of this action. Bacteria ferment the lactose (sugar) in milk, forming lactic acid. This separates the proteinous milk solid called curd from the watery, mineral-laden liquid called whey. The growth of spoilage microbes is slowed in the presence of the acid, and the curd becomes cheese.

The Chinese found centuries ago that sprouting alters seed chemistry. Sprouting changes starch into easily digested malt sugar, activating the hormone auxin, which increases the size and number of plant cells. A sprouted seed also has ten times the nutrient value of an ungerminated seed. Its vitamin content is multiplied, making sprouts one of the best sources of vitamins A, B, C, D, and E. One seven-member family lived for six months in the 1960s on a diet of sprouted seed. None of the family became ill or malnourished, and the total food cost for the period was $52.50. We caution the reader, however, to beware treated seed, which can be deadly if ingested. Check your sources carefully and buy only a chemical-free product—if purchase you must.

You can make any whole and dried bean, pea, or grain sprout in several days, without purchasing a seed sprouter to do the job. Use the common, wide-mouthed quart canning jar and allow several tablespoonfuls of any seed to soak in tepid water for a few hours. While soaking commences, punch the jar's lid with as many holes as possible, from the inside out. This will permit abundant air flow, providing the bursting seed with oxygen for nutrient combustion. After allowing the seeds to soak for a few hours, screw the jar lid in place and invert the jar, draining the mineral-laden water into a storage receptacle so that you can later use it to make soup. Place the jar of sprouts in a sunny window or let it stand in a comfortably warm room. As sprouts fill the jar they may find a bit more room, more air movement, and less compaction if the jar is turned on its side. Seed sprouts thoroughly shaken in warm water and drained morning and night will exhibit accelerated growth in a matter of days. They greatly enhance many kinds of dishes when added uncooked before serving, and children delight in watching them grow and change.

Sprouting is rehydrating; that is, moisture is added to dehydrated seeds, leaving them prone to rapid spoilage unless eaten when fully mature, before green appears in their first leaves. Grains and seeds cannot be preserved unless they are thoroughly dried after harvest. Spoilage organisms do not thrive in the absence of moisture. Sun-drying fruits and vegetables in the open has the advantage of free solar heat but the disadvantage of your being unable to control the temperature and humidity of the process. Notable nutrient losses result.

Such losses can be minimized by using a controlled-heat dehydrator which operates after sundown, through periods of weather change, and to accommodate certain late-ripen-

SEED	TEMP.	RINSE/DAY	DAYS	LENGTH	YIELD
ALFALFA	72°F	2 TIMES	3-5	1-INCH	1:10
BARLEY	70-80	2-3	3-4	½	1:2
BEANS	68-86	3-4	3-5	1-2	1:4
BUCKWHEAT	68-80	1	2-3	1	1:3
CHICK-PEAS	68-72	4-6	3	½	1:3
CORN	72-86	2-3	2-3	½	1:2
FLAX	68-78	2-3	4	1	1:5
LENTILS	68-86	2-3	3-4	½	1:6
MILLET	70-80	2-3	3-4	¼	1:2
MUNG BEANS	68-86	3-4	3-4	2	1:5
OATS	70-80	1	3-4	½	1:2
PEAS	50-68	2-3	3	½	1:1½
PUMPKIN	68-86	2-3	3	¼	1:2
RICE	50-80	2-3	3-4	½	1:2½
RYE	50-68	2-3	3-4	½	1:3½
SESAME	68-80	4+	3	½	1:1½
SOYBEANS	68-86	5	4	2	1:4
SUNFLOWER	65-75	2	2-3	½	1:3
VEGETABLES	68-86	2	3-5	1-2	1:5
WHEAT	70-80	2-3	3-4	½	1:4

23.1 SROUTING

ing foods. The key to dehydration is temperature and humidity control and air circulation. When air circulation through the drying food is inadequate, moisture evaporates too rapidly from the surface of the drying substance. This is especially true of fruits high in sugar and results in a glazed surface, which is brittle rather than rubbery, as it should be.

By controlling the amount of air exhausted from the dehydrator, you can maintain humidity in the dehydrator at a relatively high rate for about five hours. Then increase the circulation to the outside. Air velocity ceases to affect the drying rate as moisture subsides. In general, a dehydrator that uses a high volume of airflow is desirable. Air circulation *through* drying food is more effective than air circulating *over* it.

The temperature range for dehydrating vegetables is from

120 degrees to 150 degrees Fahrenheit for a period of one to eight hours. During the first few hours apply the maximum amount of heat until most of the moisture is evaporated. After this initial period of rapid evaporation reduce the temperature and increase air circulation. At the end of six hours, the water content of most fruits or vegetables will be reduced 8 to 10 percent. A five-hour period of air-drying at atmospheric (room) temperature follows. During this curing period, moisture will be further reduced to 5 percent or less. Curing prevents losses of vitamins, flavor, and texture. One study conducted on sun-drying Thompson seedless grapes determined that there is a near-total destruction of vitamin A. Grapes dehydrated in a cabinet showed no appreciable loss of this vitamin.

The dehydrator we use is part of our household wood heating and cooking system. We designed the stove as a single unit to function for multiple uses. Heating rooms, heating water, cooking, baking, smoking meat, and dehydrating are fulfilled by merely adjusting a damper, turning a switch, or inserting a baffle plate.

The stove combines the most important combustion features known for wood-burning stoves. This design has a dual preheated draft inlet for more even and complete fuel ignition. It includes a secondary combustion flue in which accumulated second-stage gases are ignited. There is a warm-air chamber from which, using a small blower fan, heat can be circulated to other rooms. The stove is a good, functional example of the principle of complete combustion. Its combustion chamber, the heart of the unit, consists of a 35-gallon oil drum set inside, and welded to, a 55-gallon oil drum, which serves as an outer shell. We chose oil drums because of their availability, low cost, and strong curvilinear construction. To provide a heat-absorbing cooking surface,

REAR AND FRONT VIEWS — DESIGN: AUTHOR — PHOTO: J. RAABE

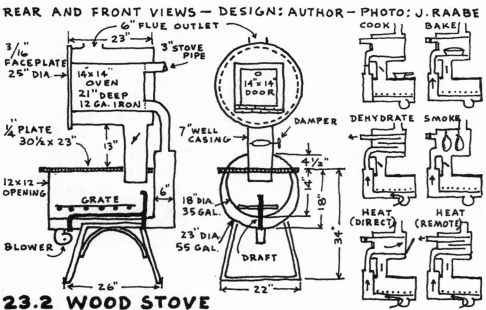

COOK | BAKE |

3/16" FACEPLATE 25" DIA.

6" FLUE OUTLET

23"

3" STOVE PIPE

14x14" OVEN 21" DEEP 12 GA. IRON

14"x14" DOOR

7" WELL CASING

DAMPER

¼" PLATE 30½ x 23"

13"

12x12 OPENING

GRATE

6"

18" DIA. 35 GAL.

4½"

4"

18"

34"

BLOWER

23" DIA. 55 GAL.

DRAFT

22"

26"

DEHYDRATE | SMOKE

HEAT (DIRECT) | HEAT (REMOTE)

23.2 WOOD STOVE

we welded a sheet of ¼-inch steel plate to the top of the intersecting drums. Spaced bars of 1-inch reinforcing steel form the grate. Below the grate a continuous length of 2-inch-diameter steel pipe is welded to a bracket which has a hinge action, allowing the grate to be elevated for cleaning out ash. Drilled holes perforate the bottom of the pipe and emit a continuous draft of air the full length of the fuel bed.

Draft inlets at the front of the stove allow for quick starting and at the rear for the complete combustion of gases accumulating there. Both draft inlets are equipped with positive, fine-adjustment controls for opening and closing. The rear draft is supplied from air space under the floor, thereby creating a full-circle, no-draft convection current. In winter we use a silent, front-mounted, squirrel-cage fan to circulate jacketed hot air into the room. A second 55-gallon drum encases the oven at a convenient level above the cooking surface. Hot air flows to rooms from the central chamber of this upper unit, which acts as a heat exchanger. One of the stove's best features is that it can be made in the homestead workshop for very little cost using readily available parts. Figure 23.2 illustrates the features of its construction and its functions.

In areas with few reliable winter food sources, cooling perishable foods adds months to their shelf life. The growth of microbes, including bacteria, yeasts, and molds, is slowed down by refrigerant cooling and stopped by freezing, which turns all the water in a food to ice. Whole carcasses of livestock butchered in winter can be air-cooled aloft at night and lowered into insulated underground pits during the day. A cooler cupboard vented at top and bottom has been a convenient cooling device for freshly picked produce for generations in our family. It operates, of course, at the coolest outside air temperature.

Insulated root cellars are traditional and worthwhile adjuncts to any serious program of food preservation. The two key factors controlling temperature and humidity in root cellars are insulation and ventilation. For this reason most cellars are located underground, where temperature varies little. Air moves upward by stack effect. Actually, after their harvest, foods continue to respire, combining carbohydrates in their cells with oxygen from the air. Water and carbon dioxide are formed and heat energy is released, the amount varying with the commodity and its storage temperature. A ton of apples, for example, stored at 32 degrees Fahrenheit will develop 1,500 BTU (British Thermal Units) in twenty-four hours. If the temperature is kept at 60 degrees, the apples will generate 7,880 BTU in the same period of time. The most perishable products are those with the highest rates of respiration. When their temperature is reduced, respiration is reduced, slowing the conversion of stored sugars and starches.

Respiration ratings have been established for many foodstuffs, and Figure 23.3 shows some of them. The chart also shows the respiratory characteristics of the more important fresh fruits and vegetables. Notice that vegetables requiring a warmer temperature in which to grow also require drier air for their storage. Cool-growing varieties, by contrast, require moist air. Ventilation is therefore essential for warm-weather crops, which are most often stored in a dry basement or attic. Crowding at any time during the process invites spoilage.

A basement food-processing and storage facility is illustrated in Figure 23.4. The root cellar's door and walls are heavily insulated. Cool incoming air enters the cellar at floor level through a 6-inch drainpipe. A food-processing sink and the basement floor both drain from this air inlet. Ventilation

23.3 FOOD STORAGE REQUIREMENTS

FOOD	STORAGE TEMP. °F	CONDITION REL. HUM.%	STORAGE LIFE	VENT. REQUIRED	RESPIRATE. MIL. BTU
APPLE	30-32	85-90	2-5 MO	HIGH	1-4
APRICOT	31-32	85-90	1-2 WK		
ASPARAGUS	32-35	90+	3-4 WK		
BEAN. GREEN	45-50	85-90	8-10 DA	HIGH	OVER 30
LIMA	32-40	85-90	10-20 DA		10-30
BEET. TOPPED	32	90-95	1-3 MO	LOW	7-10
BUNCHED	32	90-95	10-14 DA		
BROCCOLI	32	90-95	7-10 DA		OVER 30
BRUSSELS SPROUT	32	90-95	3-4 WK		
CABBAGE	32	90-95	3-4 MO	LOW	4-7
CARROT. TOPPED	32	90-95	4-5 MO	LOW	7-10
CAULIFLOWER	32	85-90	2-3 WK		
CELERY	31-32	90-95	2-4 WK		7-10
CHERRY	32	85-90	10-14 DA	LOW	10-30
CORN - SWEET	31-32	85-90	4-8 DA		OVER 30
CUCUMBER	45-50	85-90	2-3 WK		10-30
CRANBERRY	36-40	85-90	1-3 MO	LOW	1-4
ENDIVE	32	90-95	2-3 WK		
GARLIC - DRY	32	70-75	6-8 MO	HIGH	
GRAPE	30-31	85-90	3-6 MO	LOW	1-4
LETTUCE	32	90-95	2-3 WK	LOW	OVER 30
MELON	40-50	85-90	2-3 WK		7-10
ONION. DRY	32	70-75	6-8 MO	HIGH	1-4
PARSNIP	32	90-95	2-4 MO	LOW	
PEACH	31-32	85-90	2-4 WK	LOW	7-10
PEAR	30-31	90-95	2-7 MO	LOW	10-30
PEA - GREEN	32	85-90	1-2 WK	HIGH	OVER 30
PEPPER-SWEET	45-50	85-90	8-10 DA		7-10
PLUM + PRUNE	31-32	85-90	3-4 WK	LOW	
POTATOES - LATE	38-50	85-90	5-8 MO	MED	1-4
PUMPKIN	50-55	70-75	2-6 MO	HIGH	
SQUASH - WINTER	50-55	70-75	4-6 MO	HIGH	
STRAWBERRY	31-32	85-90	7-10 DA	LOW	10-30
SWEET POTATOES	55-60	85-90	4-6 MO	MED	4-7
TOMATO RIPE	50	85-90	8-12 DA	HIGH	4-7
GREEN	55-70	85-90	2-6 WK	HIGH	
TURNIP	32	90-95	4-5 MO		4-7
SPINACH	32	90-95	10-14 DA		OVER 30

is created with a revolving ventilator located on the roof of the building. It is possible to reverse air circulation in the root cellar during evening hours, distributing night sky coolness that is reflected from panels on the roof's solar unit.

From our point of view a minimum of preservation is desirable in the homestead food program. A family can adjust its diet to utilize and enjoy foods in season. Ideally, a greenhouse and garden maintain a continuous, year-round vegetable supply. In many regions crops such as beets, cabbage, carrots, kale, and parsnips can be wintered-over (un-harvested) outside in the garden under an insulating layer of snow or mulch. More animal products may be consumed in winter months, when fresh vegetable protein is less readily available. Seeds and grains, peas, beans, and sprouts provide abundant winter variety. There are many jobs to be done all the time on the homestead. We do not believe in making extra work for ourselves when, if we tune into the rhythm of the growing seasons, we can produce the climatized foods with their in-season taste experience, which so delighted Henry Thoreau.

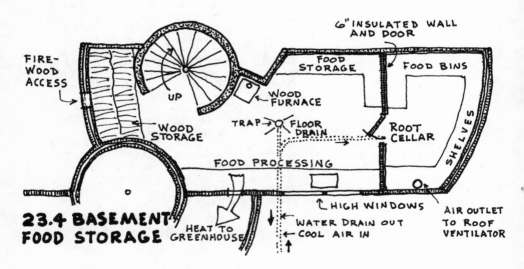

23.4 BASEMENT FOOD STORAGE

24

Nutrition

Food is one of man's most immediate points of contact with Nature. Man's sustenance must be suited to the laws that govern his body. Our ability to adapt cannot stretch beyond narrow limits.   S. *Giedion*

As an art, eating is ancient. As a science, it is very young. With the possible exception of work routine, no other discipline can have a more profound influence on the success or failure of the homestead endeavor than the choice of family diet.

As a relatively new science, nutrition is chock-full of faddism, half-truths, and ignorance. Society on this planet has yet to establish through scientific analysis all of the nutritive elements necessary to humans. To confound research in this area further, nutritional requirements vary greatly even among members of the same culture. Food preferences today continue to be based on arbitrary values, habit, and symbolic and religious tradition.

As modern life becomes more complex, the number of edible foods dwindles. For example, civilized persons now turn away from a diet of insects, yet these invertebrates remain a staple among less-civilized people living in remote areas of the globe. Dried locust cannot be matched for its 75 percent protein. Termites are far superior to animal flesh, containing 36 percent protein, 44 percent fat, and 561 calories per 100 grams. The ancient Chinese conceded that "anything that lives is edible—provided you know how to cook it."

Eating habits and food selection are subject to custom and miseducation. Apparently, when mankind became agrarian, a single lengthy eating period was arranged for each midday so that fieldwork need not be broken by frequent intermission. The preparation of this repast was done by women who subsequently rejoined the men at work in the fields. This system worked well and spread throughout the world farming community. With the rise of industrialism three meals a day were scheduled and staggered so that machines would never become idle.

Animal feeding is in large measure similar to human eating practice inasmuch as the intestinal tract is efficiently designed to handle frequent ingestion of food. A sudden influx of a large quantity of food imposes on the human digestive system more than it can utilize at one time. As large amounts are digested, the gastrointestinal tract draws blood for long periods from other parts of the body, shunting it from the main arterial system to the digestive tract. Vital organs are thus deprived of the constant supply of oxygen they need. Now that we cook many of our foods, smaller quantities and frequent nibbling are more conducive to the complete assimilation of food. Therefore, less should be consumed at one time, avoiding abnormal fat deposits.

The diet dilemma faced by conscientious homesteaders can be resolved by examining some of the diverse views offered on one hand by health-food faddists and on the other hand by establishment nutritionists with their medical bias. Both factions are correct in their first premise that chronic degenerative diseases in privileged countries are rapidly increasing the world over, although with less frequency in Third World countries. Obesity is commonplace among the privileged. Caries, cancer, heart disease, and arteriosclerosis are rampant. Half of all older adult Americans suffer diverticulitis, a severe intestinal disorder caused by the lack of bulk in the diet. The American dental surgeon Dr. Weston Price was one of the first scientists to evaluate the changes wrought in the diet of primitive people by the introduction of civilized food. He studied fifty different tribal communities throughout the world in which health and physique had been outstanding before they were influenced and corrupted by alien food selections. Price's book *Nutrition and Physical Degeneration,* from which we quote the following, is a classic.

> After spending several years approaching this problem by both clinical and laboratory research methods, I interpreted the accumulating evidence as strongly indicating the absence of some essential factors from our modern program, rather than the presence of injurious factors. This immediately indicated the need for obtaining controls. To accomplish this, it became necessary to locate immune groups which were found readily as isolated remnants of primitive racial stocks in different parts of the world. A critical examination of these groups revealed a high immunity to many of our serious affections so long as they were sufficiently isolated from our modern civilization and living in accordance with the nutritional programs which were directed by the accumulated wisdom of the group. In every instance where individuals of the same racial stocks who had lost this isolation and who had adopted the foods and food habits of our modern civilization were examined, there was an early loss of the high immunity characteristics of the isolated group and also from the displaced foods of our modern civilization.

In all fifty communities he studied, Price found that climate, diet, religion, and environment might be different but that, in the case of each group, the members grew food in fertile soil and ate it whole, directly from its source. Where the diet was influenced by civilization, there were marked changes. Dental deformities and tooth decay were common. The bacteria that produce tooth-decaying acids live on starch and sugar, the mainstays of civilized diet.

The widely varying diagnoses given by food faddists for modern physical degeneration are often extreme and ill-founded. What is there about eating the novel and unorthodox that makes this aberrant behavior popular with—for the most part—otherwise broad-minded people? Among those who in this century proclaimed the virtues of their own, widely divergent, views about food and nutrition were Eugene Debs, Upton Sinclair, and George Bernard Shaw. Past proponents of esoteric dieting were Confucius, Lao Tze, Zoroaster, and Moses. Perhaps radicalism is somehow associated with digestive discomfort or malnourishment. Actually, any form of food faddism may keep a neurotic happy. When told that blackstrap molasses, wheat germ, and brewer's yeast contributed to longer life, Jimmy Durante retorted, "Not so. They just make it *seem* longer!"

Vegetarians lead the food extremists in numbers. They reject meat and animal products as putrefactive, although some amino acids and B-complex vitamins are difficult to obtain from sources other than animals. So why restrict them? The macrobiotic diet fad flared for a time, suited to the inactivity of the contemplative life, yet it was dangerously lacking in protein. This bland, monotonous, narrow selection of foods has killed more than one devotee.

A large percentage of the American population is hooked on diet additives, vitamin and mineral tablets, and costly food supplements of all kinds. At one time in the 1950s there were 15,000 salespeople peddling a food supplement called Nutrilite, which is still sold today under another name. As successfully marketed as Heinz "57" varieties of canned food, this expensive supplement—consisting simply of a tablet of processed alfalfa, parsley, and watercress—was supposed to heal fifty-seven different human diseases. Vitamin freaks ascribe all health to the value of a bottle of pills, but as a result, overdosing of vitamins A, D, and E has become common. It is increasingly apparent to nutritionists themselves that unprescribed amounts of one vitamin may cause an increased need for another, often creating a deficiency where none existed before.

A new miracle food seems everlastingly before us. Well-known nutritionists, who have each authored numerous books favoring their own special bias, appear to be at one time or another financially involved with a company selling their chosen cure-all. Blackstrap molasses, which is high in minerals, is merely the first discard from the sugar-refining process. It has been calculated that to get one's daily allowance of iron, a person would have to consume a gallon of this molasses. Yogurt has been highly overrated; it has only about the same nutritional value as plain milk. Royal jelly was once erroneously purported to stimulate sexual appetite and prowess. Then there was the honey-vinegar cure for just about every ailment known, leaving us to snort, justifiably and cynically, "What next?"

An organization calling itself the National Health Federation has become spokesman and political protector for many of the supplement peddlers, ranging from health-food stores to hawkers of electronic gadgetry and detractors of medical intervention in critical human illness. Natural hygienists who support this organization are vegetarians who consume only raw foods and fast for bodily rehabilitation. For hygienists bodily ills are simply created by an accumulation of toxins, which require periodic elimination by withholding all food for some days. The cause of human illness, they feel, is acidosis, the simultaneous combination of protein and carbohydrate in the digestive system. They conclude that because protein requires acid for its digestion and carbohydrates require an alkaline medium, starches and protein should never be eaten at the same meal. This, however, does not account for the fact that some foods, namely peas and beans, contain both starch and protein naturally. The whole acid-alkaline, yin-yang esotericism seems invalid to us.

It is not our purpose, however, to refute the conclusion shared by all concerned with human health that something is terribly wrong with civilized peoples' health. We wholeheartedly subscribe to the contention that commercially processed, treated foods must be avoided. No preservative, artificial coloring or flavoring, modifier, bleach, emulsifier, or antioxidant should be added to our foods. Neither should the foods we consume be robbed of essential nutrients in their growing or processing. By no means should life-supportive foods be subjected to pesticides, herbicides, antibiotics, or hormone injection.

Establishment nutritionists, largely sponsored by the medical profession, see no problem with these methods. Their focus is on such obscurities as "minimum daily requirements" and calories. Three major nutrients can be identified—carbohydrate, fat, and protein—each of which humans need each day to supply physical energy for daily tasks and to provide nutrients for cell construction and regeneration. Well-being and regeneration depend on a balance of amino acids, calories, and essential fats, vitamins, and minerals. The prescribed minimum daily requirement is, however, far too high to allow individual variances. And the foods we eat differ greatly in their value, making it unthinkable to try to prescribe an international unit, calorie, or any other arbitrary measure. In an examination of a thousand people it was found that the requirement for vitamin B6, pyridoxine, varied from 5 milligrams daily to as much as 400. In a representative sampling of young men the calcium requirement varied fivefold. Often people are able to absorb only one-fifth of the calcium they receive. Vitamin D is instrumental to the absorption of calcium for bone growth but may be toxic to some people.

Differences in the composition of foods are even greater than individual variances. The vitamin-C content of tomatoes depends upon how much sunlight strikes the fruit just before harvest. Overly rich soil may produce large leaves, which shade fruit, lowering its vitamin C. In 1948 Dr. Firnan Bear reported to the Soil Science Society of America that the iron content of tomatoes varied from 1 part to 2,000 parts per million. He found that calcium in lettuce may vary 400 percent and that copper in spinach will fluctuate between 12 and 88 parts per million.

The universal balanced diet for all persons is an absurdity. People differ widely in physical stature, constitution, and nutritional requirement. The 30-odd tons of food consumed in a lifetime therefore have a decisive influence on a person's behavior.

The question finally becomes "On what basis does one make a dietary choice?" The body, for all its regenerative capacity, dies a little each day. No specialized regimen can stop this. The best we may do is try to choose a balance, a simple diet consisting of as great as possible a variety of fresh foods, least removed in time from their growing environment. It is reasonably well known that the human body has its own control mechanisms, that we need not rely on the clock or other arbitrarily set patterns to satisfy our food needs. A part of the evil of the civilized diet lies in its disordering of the control centers in the brain, which guide our selection. When foods are wholesome, the body informs the consciousness of its needs. In early experiments infants and young children unconsciously selected a balanced diet when there was a choice of whole, natural foods. They did not consciously choose a balanced diet on any given day; a child would, however, concentrate on one food for a while, perhaps for several days in a row, unconsciously supplying a specific need. Later, after indulging a particular food, the child would eat various other foods or another specific one. In time, intake balanced. It appears that when a food is balanced, the body "learns" its value, triggering an automatic bodily response. Likewise, it may be that the cravings of pregnant women are body signals that they require a particular nutrient.

It is unrealistic to seek complete physical well-being, any more than it is possible to find total peace of mind. People constantly establish and reestablish dynamic equilibrium with their environment. It is not practical to concern ourselves with every possible deficiency. We are confronted with the requirement for sixteen different minerals, seventeen essential vitamins, and ten indispensable amino acids. It may be that the individual thrives on momentary disequilibrium. We should be open to variety in our diet. Just eat an entire meal of starch, and you will soon after notice hunger signs, for the monotonous mass will soon be flushed through the digestive system. If some fat is eaten with the starch, however, it delays passage from the stomach, increasing the flow of gastric juices and slowing down the progression.

There are only two comparatively complete foods—milk and eggs. The homestead cow's raw milk will possibly be too lacking in iron to be complete, but milk from commercial dairies is guaranteed to be deficient in several vitamins lost in pasteurization unless otherwise added. People in China and Japan remain healthy with little milk, meat, or eggs. Their diets contain large quantities of vegetable protein, such as the invaluable soybean. Peanut flour, nutritionally superior to other flours, contains four times the protein, eight times the fat, and nine times the mineral content of whole-wheat flour. Also, a crop of peanuts is simpler to grow and process than grain.

Food combinations have leaped to the forefront of people's consciousness with Francis Lappe's fine little book *Diet for a Small Planet*. Consequently, we now know that a complete, high-quality protein results from combining vegetable protein (usually deficient in the amino acid lysine) and animal protein (short on methionine). This complementary synthesis of otherwise incomplete proteins provides humans with the proteins necessary for life.

Homesteading is a self-reliant way of life on the land, and wholesome food production may well be the first step toward achieving self-reliance. This momentum toward self-reliance and self-provision has been, in recent years, appropriately labeled the Green Revolution. It has been quietly moving forward as families return to the soil to raise their crops and their children. Nothing seems substantially to impede its progress. As one unsung poet puts it, "Don't look for its soldiers in the city. Most of the real ones are long since gone to their domes and gardens, with goats and chickens the day was won. You now see only plastic imitations who will starve yelling, 'What's it all mean?'—not knowing that the revolution has come and gone and was won in a patch of beans."

Indeed, we feel that the movement of mankind to return to things of the earth, the source of our being, is irresistible and undeniable. Recent space exploration has made it abundantly clear that our solar system has no other ports of entry for us. This is it; there are no other biospheres for us to muck up willfully. Are we to become like the great gorillas of Africa, who, as they approach inevitable extinction, have taken to soiling their nests? Can we draw a parallel with our own behavior as we create radical and disastrous changes in the life-protecting atmosphere, in the great oceans, on the great land masses, and in the delicately balanced food chain that feeds us, one and all?

Mankind is about to make a choice, whether it knows it must do so or not. Let us hope it's an educated choice. The material is available for those who will study it and look about themselves to consider its validity. Some unselfish human goodwill for other living things would go a long way toward ensuring the continuance of life on earth.

I do not think that any civilization can be called complete until it has progressed from sophistication to unsophistication, and made a conscious return to simplicity of thinking and living. Lin Yutang

About Books

In the bibliography of *The Owner-Built Home,* photos of a house model were used to illustrate four aspects of home-building: site and climate, materials and skills, form and function, and design and structure. In this book we have redesigned that prototype house to function specifically as a homestead residence, incorporating solar heating and cooling, a root cellar and a summer kitchen, and a privy tower that includes water storage and a wind generator. This new design has been simplified for construction by homesteaders, for we wish to correct the idea that an inexperienced, unskilled landowner cannot build an entire homestead, developing all essential facilities. The basic tool for building these component structures for the owner-built homestead is the rotating building form (see Figure 17.2). To build a complete homestead from our plans you need few tools—including a cement mixer, wheelbarrow, and this rotating building form—and a fundamental mathematical understanding of the properties and functions of the circle. We hope that this book has provided some knowledge for the use of these tools.

In a sense, books are tools. And to a large extent today's prospective homesteader must rely on obsolete—and often very dull—book-tools. The majority of the literature on homestead topics has been written for another time and, too often, another place. At least one popular publishing company has established its current success by reissuing

Kern homestead model

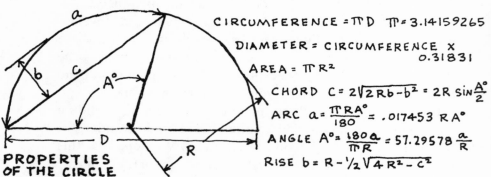

PROPERTIES OF THE CIRCLE

CIRCUMFERENCE $= \pi D$ $\pi = 3.14159265$

DIAMETER $=$ CIRCUMFERENCE $\times 0.31831$

AREA $= \pi R^2$

CHORD $C = 2\sqrt{2Rb - b^2} = 2R \sin\frac{A^\circ}{2}$

ARC $a = \frac{\pi R A^\circ}{180} = .017453 \, R A^\circ$

ANGLE $A^\circ = \frac{180 a}{\pi R} = 57.29578 \frac{a}{R}$

RISE $b = R - \frac{1}{2}\sqrt{4R^2 - C^2}$

depression era farming books, dressed in shiny new jackets to confuse the unsuspecting reader further. Other companies specialize in republishing state and federal government agricultural bulletins originally prepared years ago for the business-farming audience. And there are those authors who have rephrased the homestead literature of other countries for use in our own, different, climate zone. Many British farming and gardening manuals are distributed in this country. The English, of course, lead the world in what they call the "smallholder" movement, but their mini-farm agricultural practices are not entirely appropriate to our climate and our life-style.

The recent Dover reprint of M. G. Kains's *Five Acres and Independence* is a classic work written for the depression era. Every prospective homesteader should read it, not for its practical information, which happens to be now largely obsolete, but for Kains's clear expression of the historical background and the philosophy of homesteading. Other inspirational but dated works are Ralph Borsodi's *Flight from the City,* originally published in the thirties, and Helen and Scott Nearing's *Living the Good Life,* reprinted by Schocken Books in 1970.

John and Sally Seymour's *Farming for Self-Sufficiency* (Random House, 1970) is an excellent presentation of the British smallholder's approach to homesteading, the present homesteading movement having had its origins in Britain. During the depression, while North American farmers were battling floods and dust bowls, British agriculturists were formulating advanced concepts of soil management and plant growth. The Rodale Press reprint of Sir Albert Howard's *An Agricultural Testament* relates these early beginnings. Howard also wrote the great inspirational classic *The Soil and Health,* published by Devin-Adair in 1947.

In our introduction we speak of some of homesteading's early beginnings in the seventeenth and eighteenth centuries. The two best books for this background are Peter Kropotkin's *Field, Factories, and Workshops,* republished by Harper & Row (1974); and E. F. Schumacher's *Small Is*

Beautiful, from Sphere Books (London, 1974). We also quote from John Gifford's *The Tropical Subsistence Homestead,* printed by Colonial Press in 1934. In 1972 the University of Miami Press released a book about Gifford's work called *On Preserving Tropical Florida.* Gifford also wrote *Living by the Land,* published by Glade House (Miami, 1945), which is still available in many libraries. Finally, we urge our readers to broaden their understanding of the deplorable situation for land tenure in this country and of the stranglehold that agribusiness has on food usage by reading *The Peoples' Land,* edited by Peter Barnes and published by Rodale Press (1975).

The volumes quoted in Chapters 1 and 2 may be of little interest in general to the reading homesteader, but they are listed here for perusal by the serious student: F. Matthias Alexander's *Use of the Self* and Eric Gill's *Sacred and Secular,* published by J. M. Dent (London, 1940).

A half-dozen titles for locating country property are currently marketed, but we can recommend only one: Les Scher's *Finding and Buying Your Place in the Country* (Macmillan, 1974). A better investment would be the 75¢ purchase of a topographic map of your site, obtainable by writing to the U.S. Geological Survey Map Information Office, Washington, D.C. 20242. The necessary tools to do your own land survey can be obtained from Forestry Suppliers, Box 8397, 205 W. Rankin St., Jackson, ME 39204.

We have discovered two exceptional books about water development: *The Water Well Manual,* by Ulric Gibson for Premier Press (Berkeley, 1971); and the World Health Organization's publication from Geneva, *Water Supply for Rural Areas* (1959).

The claim to having written *the* classic work on soils must be shared by both American and British scientists. The undisputed authority from this country is William Albrecht, whose *Soil Fertility and Animal Health* was published by Fred Hahne in 1958. *Acres USA,* the energetic and progressive monthy newspaper-voice for eco-agriculture in this country, has recently compiled and published *The Albrecht Papers,* a multivolume work that should be required reading for any soil study. Selman Waksman's work *Humus,* published by John Wiley (1952), should also be required reading for this subject. The major British work on soil, *Soil Conditions and Plant Growth,* comes from the Rothamsted Experimental Station and was written early in this century by its director, E. J. Russell. A new 1973 edition has just been released from the station. Faber Publishers in Britain are well known for their many fine gardening and farming books, among them *Gardening without Digging* by A. Guest (1948), *The Earth's Green Carpet* by Louise Howard (1947), *The Living Soil* by Lady Eve Balfour (1975), and *Natural Gardens* by J. and B. Maunsell (1958).

Traditional ley farming originated in Britain, so there is little wonder that the best sod crop books have come from that source. A. T. Semple's *Grassland Improvement* from Leonard Hill (London, 1970) has been our best reference for this subject, and we highly recommend it for study. Other books that similarly treat grasses as pasture and livestock feed include André Voisin's *Grass Productivity* (1958) and his *Rational Grazing* (1962), both published by Crosly Lockwood; and Michael Graham's *Soil and Sense* (1941) and George Henderson's *The Farming Manual* (1960), both from Faber Publishers.

Richard St. Barbe Baker is quoted in our chapter on forage. His books, *Sahara Challenge* (1954) and *Green Glory* (1958), from Lutterworth Press, have much to do also with woodlands and the reforestation of deserts. For specific crops, such as comfrey, we recommend Lawrence Hill's *Comfrey*

Report, published in 1974 by the Henry Doubleday Research Association. Other excellent Faber publications include Friend Sykes's *Humus and the Farmer* (1946) and his *Modern Humus Farming* (1959), and Newman Turner's *Fertility Farming* (1951).

F. H. King is one of our most important mentors. In our chapter on row crops we refer to his inspiring book *Farmers of Forty Centuries,* recently republished by Rodale Press. And even though Ed Faulkner's three books were written in the forties, his message is still timely. *Plowman's Folly, A Second Look,* and *Soil Restoration* were all published by the University of Oklahoma Press. Other books recommended for reading about row crops include:

Joseph Cocannouer, *Weeds, Guardians of the Soil* and *Farming with Nature,* from Devin-Adair.

F. H. Hainsworth, *Agriculture: A New Approach,* Faber, 1954.

Beatrice Trum Hunter, *Gardening without Poisons,* Houghton-Mifflin, 1964.

John Jeavons, *How to Raise More Fruits and Vegetables Than You Thought Possible by the Organic Method,* Mid-Peninsula Ecology Action Group, Palo Alto, CA.

F. C. King, *The Weed Problem,* Faber, 1951.

Ehrenfreid Pfeiffer, *The Pfeiffer Garden Book,* from the Bio Dynamic Farming and Gardening Association, Stroudsburg, PA, 1967; and *The Earth's Face and Human Destiny* from Rodale Press, 1947.

The classic book on tree crops is J. Russel Smith's *Tree Crops: A Permanent Agriculture,* published by Devin-Adair in 1953. For a modern update of the same subject we recommend Douglas Sholto's *Forest Farming* from Watkins (London, 1976). Downing's *Fruit and Fruit Trees of*

America, first published in 1860, can still be found in many large libraries. D. Macer Wright's book *Fruit Trees and the Soil,* from Faber, is a more contemporary introduction to these crops.

For further reading on woodlands we suggest:

Helen and Scott Nearing, *The Maple Sugar Book,* Schocken Books, 1970.

M. C. Raynes, *Problems in Tree Nutrition,* Faber, 1944.

Steve Ross, *The Chain Saw Book,* Oliver Press, Willits, CA, 1976.

Until our own book on greenhouse designs and management (now in preparation) appears, we recommend *Organic Gardening under Glass* by Doc Abrahams (Rodale Press, 1975). John Ott's *Health and Light* from Devin-Adair (1973) is essential reading on this subject, but General A. J. Pleasonton's 1861 study, *The Influence of the Blue Rays of the Sunlight and the Blue Colour of the Sky,* is now difficult to find even in university libraries.

The Village Technology Handbook, published in 1970 by VITA (Mt. Rainier, MD), contains many practical road-building ideas. In our chapter on shop-building and working we quote Milton Wend's *How to Live in the Country without Farming,* a book published by Doubleday and so popular since its first printing in 1945 that even library copies are difficult to find. A more recent Dover reprint of *Tools and Their Uses,* originally published in 1971 by the Bureau of Naval Personnel, is a handy reference for this area of work.

Two exceptional books on animals have recently been published by Rodale Press: Frank Ashbrook's *Butchering, Processing and Preservation of Meat* and Jerome Belanger's *Homesteader's Handbook to Raising Small Livestock.* Frank Morrison's enduring work *Feeds and Feeding,* originally published in the thirties, is also published now by Rodale Press. We quoted in Chapter 19 on animal feed from *Observations on the Goat,* originally published in 1970 by the Food and Agriculture Organization of the United Nations.

A good book on small-scale homestead fish production has yet to be written. Those interested in this aspect of homesteading should read the proceedings of the United Nations Symposium of 1966 on fish culture and the FAO bulletin of 1966, "Fish Culture in Central East Africa." The book *Aquaculture,* by Bardach, Ryther, and McLarney, provides a thorough introduction to this activity.

Alternative sanitation has been the subject of a number of excellent books in recent years. The current 1975 edition of *Stop the Five-Gallon Flush!,* published by the Minimum-Cost Housing Group (School of Architecture, McGill University, Montreal, Canada), heads the list. The World Health Organization's *Excreta Disposal for Rural Areas,* written by Wagner and Lanoix in 1958, remains the authoritative work on composting privies. Other good books on the subject are Clarence Golueke's *Composting,* published by Rodale Press; and Harold Gotaas's *Composting,* published by the World Health Organization.

The all-purpose barrel stove featured in our chapter on food preservation has been described in greater detail by Bob Ross in his thorough book *Woodburning Stoves,* from Overlook Press (1976). Wood-using enthusiasts should also read Larry Gay's *Heating with Wood,* printed by Garden Way (1974). Other books related to food preservation include:

Dr. Michael Jacobsen, *Nutritional Scoreboard.*
Bruford Reynolds, *How to Survive with Sprouting,* Hawkes Publishing, 1973.
Carol Stoner, *Stocking Up,* Rodale Press, 1974.

In our final chapter we quote Dr. Weston Price's great classic, *Nutritional and Physical Degeneration,* published by the author in 1945. Other books recommended for general reading on the subject of nutrition are:

Adelle Davis, *Let's Eat Right to Keep Fit,* Harcourt Brace Jovanovich, 1970.
Ellen B. Ewald, *Recipes for a Small Planet,* Ballantine Books, 1975.
Handbook of the Nutritional Contents of Foods, #8, USDA, 1963.
Francis Lappe, *Diet for a Small Planet,* Ballantine Books, rev. ed., 1975.
Dr. Lionel Picton, *Nutrition and the Soil,* Devin-Adair, 1949.

Our next book, now in preparation for Scribners, should fill the gap in the homesteader's need for a single convenient volume of information on food production. Since the late 1800s the USDA has yearly published similar reference material. *The Small Holder's Encyclopedia* was published in Britain in 1950 and Rodale Press compiled an *Encyclopedia of Organic Gardening* in 1959. At least six volumes of the USDA *Yearbook of Agriculture* remain pertinent today: *Better Plants and Animals* (1937), *Soils and Men* (1938), *Climate and Man* (1941), *Grass* (1948), *Trees* (1949), and *Water* (1955).

Although there is no single periodical specifically directed to the homesteader, much can be gleaned from a number of current publications:

Acres USA, $5.00/year, 10227 E. 61st St., Raytown, MO 64133.

Alternative Sources of Energy Newsletter, $6.00/year, Rt. 1, Box 90-A, Milaca, MN 56353.

Biodynamics Quarterly, RD 1, Stroudsburg, PA 18360.

Countryside, $9.00/year, 312 Portland Rd., Highway 19 East, Waterloo, WI 53594.

The Mother Earth News, $10.00/year, P.O. Box 70, Hendersonville, NC 28739.

Organic Gardening and Farming Magazine, Rodale Press, Emmaus, PA 18049.

Rain, $10.00/year, 2270 NW Irving, Portland, OR 97210.

Undercurrents, $7.50/year, Undercurrents, Ltd., 274 Finchley Rd., London NW3, England.

Glossary

It is important for the fledgling homesteader to have a clear and concise understanding of common agricultural expressions. You will find numerous advantages, both social and economic, in being able to communicate intelligently with neighboring farm folk, as well as with fellow homesteaders. You will also find that proper word usage is especially important to these people who earn their livelihood by producing foodstuffs; to call a heifer a cow, or a steer a bull, can be an unforgivable breach of common knowledge in the eyes of those who work with these critters daily. The following chart summarizes the terms commonly applied to some primary farm animals.

	CATTLE	GOAT	SHEEP	SWINE	POULTRY
Genus	*Bos*	*Capra*	*Gallus*	*Sus*	*Ovis*
Groups	Herd	Band	Flock	Drove	Flock
Newborn	Calf	Kid	Lamb	Pig	Chick
Young male	Bullock	Buck kid	Ram lamb	Shoat	Chick
Young female	Heifer	Doe kid	Ewe lamb	Gilt	Chick
Male of breeding age	Bull	Buck	Ram	Boar	Cock
Mature female	Cow	Doe	Ewe	Sow	Hen
Unsexed male	Steer	Wether	Wether	Barrow	Capon
Unsexed female	Spayed	Spayed	Spayed	Spayed	

adaptation. A change in structure, function, or form that enables an animal or plant to adjust to its environment.

aeration or *airing.* Exposing soil to the air by plowing, harrowing, etc.

aerobic bacteria. Bacteria that require free oxygen to live and function.

aftermath pasture. The second growth of a previously harvested field on which animals are turned to graze.

aggregate. An agglomerate of different minerals that are separable by mechanical means. Also, sand and pebbles added to cement to make concrete.

agriculture. The utilization of biological processes to produce food and other products useful to people.

agrology. The science of soils as they relate to crop production.

agronomy. The art and science of crop production; the management of farmland.

algae. Chlorophyll-bearing plants such as seaweeds and pond scums that have no true roots, stems, or leaves. Found in water or damp places, they manufacture food by photosynthesis.

alluvium. Mud, sand, and silt deposited by the action of waves or currents of water.

amino acids. A group of nitrogenous organic compounds that serve as structural units of proteins and are essential to human metabolism.

anaerobic bacteria. Microscopic organisms able to live without air or free oxygen.

animal unit. A unit represented by one mature cow or horse, equivalent to two heifers, seven sheep, or one hundred laying hens.

annual. A plant that completes its entire life cycle within a year.

apiary. A group of beehives tended for their honey.

aquaculture. The raising of plants or animals in a lake, river, or other body of water.

arable. Refers to land suitable for tillage.

artesian well. A perpendicularly bored well in which the underground water pressure is great enough to force water to the surface.

bacteria. Microscopic one-celled organisms that have no chlorophyll and multiply by simple cell division.

balanced ration. The food ration furnishing an animal its correct proportion of digestible protein, carbohydrate, and fat.

bark. The outermost protective covering of trees and some plants.

barren. Incapable of producing offspring, seed, fruit, or crops.

basin listing. Tillage in which lister furrows are dammed at regular intervals to create small basins.

battery. A series of pens, cages, etc.

bedding. Litter, such as straw or hay, used to bed animals.

biennial. A plant that forms roots and leaves the first year, produces fruit and flowers the second year, and then perishes.

biodynamic farming. The use of organic manures and compost rather than chemical fertilizer for growing crops.

biotype. A group of plants or animals with similar hereditary characteristics.

blanch. To whiten or bleach (endive or celery, for example) by keeping the leaves or stalks of the plant from the light, to improve the flavor and texture.

blight. Any atmospheric or soil condition, parasite, or insect that kills, withers, or checks the growth of plants.

bolt. To produce seed before the natural time.

branch. A lateral stem.

breed. Animals having a common origin and characteristics that distinguish them from other groups within the same species.

bridge grafting. Grafting accomplished by bridging the girdled trunk of an orchard tree using scions.

broadcast. To scatter seed, as opposed to sowing it in rows.

broodiness. The instinct of birds to sit on and cover eggs for the purpose of warming and hatching them.

browsing. Feeding by livestock on pasture grasses or the tender green branches or shoots of shrubs or trees.

bud. A small swelling or projection on a plant from which a shoot, a cluster of leaves, or a flower develops.

budding. The insertion of a bud of one tree under the bark of another for propagation; a form of grafting.

bunch grass. Any of various pasture grasses that usually grow in clumps.

calf crop. The calves produced by a herd of cattle in one season.

calorie. The unit of heat required to raise the temperature of one gram of water one degree Centigrade; used as the unit for measuring the energy produced by food when it is oxidized in the body.

cambium. The thin, mucilaginous cellular layer of tissue between the bark and the wood of woody plants from which new wood and bark develop.

capillary water. Water held in the spaces between soil particles.

carbohydrates. Organic compounds containing carbon combined with hydrogen and oxygen, including starches, sugars, and cellulose.

carrying capacity. The number of animals an acre of pastureland can support.

catch crop. A supplementary crop grown when the ground would ordinarily lie fallow, such as between plantings of two principal crops.

cereal. Any edible grain, such as wheat, oats, barley, or rye, or any grass producing such grain.

chaff. The husk or outer covering of grain that is removed by threshing or winnowing.

chisel. A tractor-drawn implement, with points about twelve inches apart, used to till the soil twelve to eighteen inches deep.

cistern. A large receptacle for storing water or other liquids.

clabber. Thick, sour milk, curdled or coagulated.

clay. A firm, plastic, fine-grained earth produced by the deposit of fine rock particles in water.

clay pan. A layer of stiff, compact, and relatively impervious clay.

climate. The total, long-time characteristics of weather of any region, including rainfall, temperature, humidity, wind direction and velocity.

cloche. A bell-shaped glass or plastic jar used to cover delicate plants.

clutch. A nest of eggs laid by a hen on consecutive days.

coat. A natural outer covering of an animal, such as skin, fur, or wool.

cob. The woody spike of an ear of corn around which the kernels grow.

cold frame. A plant-forcing structure deriving its total heat from the sun's heat, which is retained by the glass cover.

colloids. Noncrystalline particles, often gelatinous, that diffuse slowly or not at all.

colter or *coulter.* An iron blade or disc on the beam of a plow, used to cut the sod.

combine. A harvesting machine that reaps and threshes grain crops while moving over the field, leaving the straw standing.

companion crop. A secondary crop planted to increase or hasten returns on a plot of land, such as lettuce between tomatoes or clover between oats.

concentrate. Any feeding stuff relatively rich in nutrients and low in crude fiber, such as cereal grain, soybean-oil meal, cottonseed meal, cowpeas, or gluten meal.

conifer. Any tree or shrub of the cone-bearing pine family.

continuous cropping. Repeated planting of the same crop on the same land for a number of years.

contour cultivation. Cultivation of crops around hills at the same elevation.

cool-season grass. Grass that grows most luxuriantly during the cool season of the year.

cordon. A normally branching plant restricted to a single stem, commonly applied to fruit trees and soft fruit.

corral. A pen or enclosure for confining livestock.

cover crop. A crop, such as rye or clover, planted to protect the soil in winter or to fix nitrogen in the soil. When turned over it becomes a green manure crop.

cradle scythe. An attachment with wooden, fingerlike rods that catch grain straws cut by the blades of the scythe.

creep. An enclosure fenced to allow young unweaned animals to pass through to obtain special feed or pasturage but designed to exclude mature animals.

crop residue. The part of a crop that is not harvested, such as straw, cornstalks, or soybean hulls.

crop rotation. Cultivation of a succession of different crops on the same land.

crossbreeding. To breed or mix two varieties of the same species, producing a hybrid or mongrel.

cucurbit. Tendril-bearing vines producing gourdlike vegetables, such as the cucumber, squash, pumpkin, or melon.

culling. The practice of selecting or picking out rejects.

cultipacter. A compaction tool with corrugated roller.

cutting. A portion of a leaf, stem, or root of a plant capable of sending out roots for propagation.

damping-off. A diseased condition affecting seedlings or cuttings caused by parasitic fungi that invade plant tissues near the ground, producing rotting, usually with moist lesions on the stem.

deferred grazing. Withholding animals from pasture for a period of time beyond the normal beginning of the grazing season.

defoliate. Deprived of leaves, as by their natural fall.

dehydrate. To remove water by drying for the preservation of food.

denitrification. The reduction or freeing of nitrogen compounds, such as the nitrates or nitrites in soil, to ammonia, oxides of nitrogen, and free nitrogen—resulting in the escape of nitrogen into the air.

determinate growth. Growth that proceeds only during the first part of the vegetative season or period and then ceases, as in the shoots of many trees.

dew. Moisture condensed from the atmosphere in small drops upon the surfaces of cool bodies.

dibbler. A machine having two wheels with long, rounded projections on their rims that make spaced holes in soil into which seeds or seedlings may be deposited.

dioecious. Having the male reproductive organs in one individual, the female in another. In seed plants, having staminate and pistillate flowers borne on different individuals.

disc harrow. An implement for breaking up soil with revolving discs, concave, circular, edged tools of hardened steel.

distill. To extract certain products or to purify substances by the evaporation and condensation of the vapor of that substance, liquid or solid.

diversion terrace. A fairly high, thrown-up row of earth with a ditch on its uphill side.

domesticate. To reclaim an animal or plant from a wild state, to bring its growth and propagation under control, and to improve it for the advantage and purposes of man through careful selection.

drill. To plant in rows; to sow, as seeds, by dribbling them along a furrow or in a row.

dry farming. Producing crops without irrigation in regions of low or otherwise unfavorable rainfall, using moisture-conserving tillage methods and drought-enduring crops.

dry rot. The decay of either seasoned or standing timber caused by fungi.

early maturing. Refers to varieties of plants with the capacity for complete development earlier than the regular season for those plants.

ecology. The branch of biology that deals with the mutual relationships between organisms and their environments.

embryo. In plants, the life germ of seeds. In animals, the unborn young.

ensilage. The process of preserving fodder (silage) in a silo.

enzymes. Organic substances that accelerate (catalyze) specific chemical transformations of materials in plants and animals, such as fermentation or oxidation.

epidermis. In seed plants, the thin layer of cells forming the external integument. In animals, the overskin or outer epithelial layer of the skin.

erosion. The wearing away of land by wind and water.

espalier. A trellis or horizontal railing on which fruit trees or shrubs are trained to grow flat in areas of limited space.

estrus. The recurrent, restricted period of sexual receptivity (also called "heat") in female animals.

eugenics. The science that deals with improving the hereditary qualities of plants or animals.

evergreen. *See* perennial.

fallow. Cropland left idle to render it mellow, destroy weeds and insects, and restore its productivity.

family. A group of closely related plants or animals.

feed conversion or *feed efficiency.* The unit of feed produced per unit of feed consumed.

feeder. A young animal that adds weight economically but does not have a high finish (fatness).

fertilization (soil). The application of nutrients to the land to promote the growth of plants.

fiber crops. Plants grown for their fibrous quality. Used in making textiles, such as cotton or flax.

fibrous-rooted. Having numerous small roots without a main or tuberous taproot.

field crops. Any agricultural crop, such as hay, grain, or cotton, grown in large fields, as distinguished from crops grown in gardens.

finish. The fat distributed in animal tissue before marketing. Also, to put animals in the best possible condition for the market, as by special feeding.

flail. To thresh by beating grain from the ear.

fodder. Coarse food for livestock that is harvested whole and cured in an erect position, such as corn, sorghum, hay, or vegetables.

forage. Vegetable food of any kind for domestic animals, such as pasturage, browse, or mast.

forcing. Hastening growth in plants at an unnatural time by artificial means.

freshen. In cattle, to come into milk or to calve.

friable. Refers to soil that is easily crumbled or reduced to small pieces.

fruit. The developed ovary of a seed plant, such as a pea pod, nut, or tomato.

fungicides. Preparations used to kill fungi on plants.

fungus. Plant lacking chlorophyll and obtaining food supplies either from living plants and animals or from decaying matter.

furrow. A trench in the earth made by a plow.

germination. The resumption of growth by the embryo in a seed after planting.

glean. To gather the leavings of a grain crop after reaping.

grade animal. An animal with both purebred and scrub parentage.

graded. In stock breeding, to improve by crossing an animal with a better breed.

grafting. The process of inserting a piece of a plant (a scion) into stock rooted in the ground so that a permanent union is effected between the two plants.

grain. The seedlike fruit of any cereal grass.

grass. Green herbage affording food for grazing animals.

green manure. An herbaceous crop, such as clover or vetch, plowed under while green for soil enrichment.

greenhouse. A glass house used to protect or cultivate tender plants grown out of season.

ground water. Water within the earth, below the unsaturated zone of percolation and above the region where all openings are closed by pressure. Its upper surface may coincide with the surface or be deep below.

hardening off. To inure plants to cold or otherwise unfavorable environmental conditions by gradual exposure to lower temperatures or by decreasing the water supply.

hardpan. A cemented or compacted layer in soils through which it is difficult to dig or excavate, resulting from the accumulation of cementing material. May be caused by continuous plowing at the same depth.

harrow. A cultivating implement having metal pegs or teeth used to pulverize and smooth soil. Also used for mulching, covering seed, removing weeds.

hatch. To bring forth young from the egg by natural or artificial incubation.

hay. The cured, dry forage of the finer-stemmed crops.

haylage. Silage that is low in moisture; grass and legume crops cut and wilted in the field to lower the moisture level for ensiling, but not sufficiently dry for baling.

hayloft. A loft or scaffold in a barn for storing hay.

heavy soil. Soil that is hard to break and prepare for plants.

heel in. To cover the roots of a plant temporarily with soil, usually used on nursery trees and shrubs.

herb. A flowering plant having certain aromatic or medicinal qualities.

herbicide. Any preparation or agent used to kill or destroy weeds.

high grade. An animal of mixed breeding in which the blood of the purebred predominates.

hogging off. Utilizing crops by allowing hogs to feed on them in the field.

hormone. An internal, organic secretion of the cells of one part, carried by the body fluid or sap of an organism to produce (excite) a specific effect on the activity of cells remote from its source.

horticulture. The science of growing fruits, vegetables, and flowers or ornamental plants.

hot bed. A bed of rich earth heated by an underlayer of manure and covered with glass.

humus. Partially decayed vegetable matter forming the organic part of the soil.

husbandry. Farming.

hybrid. The product of the union of the male and female of two diverse species.

hydrophyte. A plant that lives and grows in water or in wet soil.

hydroponics. The science of growing plants without soil in solutions containing the necessary minerals.

illuviation. The geological process by which clay is added to the subsoil.

immature soil. Unweathered and recently formed soil having a profile with definite layers.

inoculation. Implanting certain microorganisms to improve soils. Also, applying nodule bacteria to legume seeds before planting them.

insecticide. An agent or preparation for destroying insects.

intertilling. Intercropping and/or cultivating between the rows of a crop.

jerk. To cut meat into long, thin strips for drying in the sun.

kernel. A whole grain or seed of a cereal, as of wheat or corn.

lactation period. The number of days an animal secretes milk and suckles young.

larva. Immature, wingless, and often wormlike form in which certain insects hatch from the egg, and in which they remain and increase in size until the next stage.

lateral. A side shoot on any kind of tree branch.

leaching. The filtering action of excessive water which dissolves and washes plant food from topsoil.

leader. A shoot that terminates the end of a branch and will continue it in the same direction of growth during another season.

legumes. Pod-producing plants with roots bearing nodules containing bacteria with the ability to assimilate nitrogen from the air.

limestone. Natural deposits of calcium carbonate or magnesium carbonate (called dolomite).

list. A small ridge of earth formed between two furrows.

lister. An implement used for furrowing land, frequently bearing a seed drill attachment.

litter. The young born at one time of any animal normally producing several offspring. Also, straw or similar material used as bedding for animals, as a protective covering for plants, or as scratch material for fowl.

livestock. Domestic animals kept for use on a farm or raised for sale and profit, usually not including poultry.

loam. A rich, dark soil mixture of equal proportions of sand and clay, containing some organic matter.

manure. Any substance put on or into the soil to fertilize it, such as animal excrement, guano, or compost.

marbling. The intermixture of fat and lean muscular tissue giving meat a veined, variegated appearance.

marginal land. Land that is too unproductive for profitable farming.

mattock. A grubbing hoe for loosening soil and digging up and cutting roots.

mature soil. Soil with a well-developed profile, having compact subsoil horizons containing distinct clay accumulations.

maverick. An unbranded animal running at large.

meadow. A tract of grassland, especially one producing grass suitable for hay.

mellow soil. A soft, porous, granular soil that pulverizes easily for good seedbedding.

mesophyte. Any plant adapted to grow under conditions of medium moisture.

metabolism. The process by which living cells transform nutrients into protoplasm and by which protoplasm is used and broken down into simpler substances and waste matter, with the release of energy for all vital processes.

millet. Cereal grasses whose small grain is used for food or forage.

mold. A downy or furry growth on the surface of organic matter, caused by fungi, especially in the presence of dampness or decay.

moldboard. The curved iron plate of a plow which receives the soil that has been cut loose and turns it over.

molting. The shedding by certain animals (such as reptiles and birds) of feathers, hair, skin, horns, etc., at certain intervals, before replacement of the cast-off parts by new growth.

mow. To cut grass or grain with a mower or scythe.

mud-capping. Placing explosives atop an object to be blasted and covering the whole with mud.

mulch. A covering or top-dressing—usually of loose material such as leaves or straw—spread on the ground around plants to prevent evaporation of water from soil or the freezing of roots.

muley. A polled or hornless cow.

mycosis. The growth of parasitic fungi in any part of the body.

native pasture. Pasture composed of native or exotic plants that have become naturalized.

neutral soil. Soil with a pH of 7.0, neither acid nor alkaline.

Newcastle's disease. Respiratory distress common to poultry, leading to paralysis and death.

nitrification or *nitrifying.* To cause the oxidation of ammonium salts, etc., to nitrites and nitrates by the action of soil bacteria.

noxious. A term applied to pestiferous, harmful weeds.

nurse cow. A milk cow used to supply milk for nursing calves other than her own.

nurse crops. Companion crops used to protect another crop sown with them from insect infestation, such as clover with grain, alfalfa with cotton.

nutriment. A substance that promotes the growth of plants or animals.

offshoot. A shoot or stem growing laterally from the main stem of a plant.

on the hoof. The designation given to a living meat animal.

organic. Refers to chemical compounds containing carbon as an essential ingredient.

outcrop. To appear at the surface of the ground, as a strata of rock.

oxidation. The union of a substance with oxygen.

pan. A hard, horizontal layer of soil, impermeable to air and moisture.

parasite. An organism that lives on or within another (host) and from which it obtains nourishment or protection without making compensation.

pasture. A crop of grass or other growing plant harvested by grazing livestock.

pathogens. Any microscopic organisms or viruses that can cause disease.

peat. Partly decayed, moisture-absorbing organic plant matter found in ancient bogs and swamps, used as a plant covering or for fuel.

perennial. A plant that lives more than two years, whether it retains its leaves or not. One that retains its leaves during the winter is called *evergreen;* one that casts off its leaves is called *deciduous.*

permanent pasture. Pasture used indefinitely for grazing by livestock.

pH. Parts of hydrogen, the symbol used as a scale to express the acidity or alkalinity of soil.

photoperiod. Light period or length of day.

photosynthesis. A natural process in living plants by which water and carbon dioxide combine to form sugars and starches (carbohydrates) by the action of sunlight on the chlorophyll in plant leaves.

pickle. To preserve or flavor food in a brine, vinegar, or spicy solution.

pisciculture. The breeding and rearing of fish.

pit. A frame-covered excavation in the soil for protecting plants.

pitchfork. A large, long-handled fork with from two to five tines used to throw hay or sheaves of grain.

pitting. The process of making shallow pits in the soil for water retention.

plankton. Microscopic plant and animal life found passively floating or drifting in a body of water, used as food by fish.

plow. A farm implement for turning up, breaking, and preparing the ground for receiving seeds.

pod. A seed vessel of plants that dries and opens when ripe.

polled. Being without horns or antlers, or having them removed.

pollen. The male element in the fertilization of a flower.

pollination. The conveyance of pollen from the anther to the stigma of a flower.

pomace. Crushed pulp from any substance from which the juices or oils have been extracted.

pomiculture. The cultivation of tree fruit.

prairie. A large, treeless tract, level or slightly rolling, highly fertile grassland, especially such an area in the Mississippi Valley.

preservative. Any agent that prevents decay or injury.

propagate. To multiply or reproduce.

protein. Any of a class of nitrogenous substances consisting of a complex union of amino acids and containing carbon, hydrogen, nitrogen, oxygen, and, frequently, sulfur; proteins occur in all animal and vegetable matter and are essential to the diet of animals.

prune. To remove dead or living parts from a plant to increase fruit or flower production or improve the form.

pseudo-cereals. Plants with many of the characteristics of cereals but not true cereals, such as buckwheat, millet, sorghum.

pulse. The edible seeds of peas, beans, lentils, and similar plants having pods.

purebred. Belonging to a breed with recognized characteristics maintained through generations of unmixed descent.

rake. Any of various long-handled farm tools with teeth or prongs at one end, used for gathering loose hay, leaves, etc., or for smoothing broken ground.

rangelands. Large, open areas of natural pastureland in the western United States over which livestock can wander and graze.

ration. The fixed portion or food allowance for an animal in a twenty-four-hour period.

reaping machine. A machine for harvesting standing grain.

refined foods. Food products from which some (or all) of the minerals and vitamins have been taken during processing, such as white flour or refined sugar.

regurgitate. To return small amounts of partly digested food from the stomach to the mouth, as in ruminant animals.

relative humidity. The ratio of the weight of water vapor contained in a given volume of air to the weight the same volume of air would hold when saturated.

repellents. Materials, such as lime or oil, used to drive away plant pests or other insects.

resin. A hardened or semisolid organic secretion exuded from certain trees and plants.

respiration. The process by which a living organism or cell takes in oxygen from the air or water, distributes and utilizes it in oxidation, and gives off products of oxidation, especially carbon dioxide.

rick. A rounded pile or stack of hay or grain, generally thatched or covered to preserve it from rain.

riparian rights. The legal rights regarding a waterway that belongs to one who owns land bordering it.

roost. A perch on which birds, especially domestic fowl, can rest or sleep.

root. A part of the body of a plant that grows downward into the soil, holding it in position, drawing nourishment and water from the soil, and storing food.

root crops. Crops grown for their single, enlarged root, such as turnips, parsnips, rutabagas, beets, carrots.

root hairs. The dense, thin-walled, hairlike, tubular outgrowths a short distance back from the tip of the growing end of a main root, which absorb water and minerals from the soil.

rootstock. A root, a rootlike stem, or a branch used as stock in the propagation of plants.

roughage. Coarse food or fodder, such as bran, straw, or vegetable peel, containing a relatively high proportion of cellulose and other indigestible constituents and serving in the diet as a stimulus to peristalsis.

rowen. Second growth of grass or hay after harvest; a stubble field not plowed until autumn, so that it may be cropped by cattle.

ruminant. A cud-chewing animal having four stomachs with four complete cavities.

sand. Loose, gritty particles of worn or disintegrated rock, finer than gravel and coarser than dust.

sapling. A young tree.

saturated fat. A completely hydrogenated fat.

scion. A shoot or twig, especially one taken for the purpose of being grafted upon another tree or for planting.

scours. A condition of persistent diarrhea in an animal.

scrub animal. An animal inferior in quality and breed, one with no purebred close relatives.

scythe. An instrument used in mowing or reaping, consisting of a long, curving blade with a sharp edge fastened to a handle, or snath, which is bent into a convenient form for swinging the blade to advantage.

seed. A fertilized and mature ovule of a flowering plant; an embryo plant supplied with sufficient food to develop into a new plant.

seedling. A very young plant grown from a seed, as distinguished from one propagated by layers, buds, etc.

seedling pasture. A field of plants in which animals are placed to graze on the young growth therein.

selective cutting. A practice in forestry of cutting out trees that interfere with the proper growth of any remaining trees.

self-feeder. Any feeding device from which animals may eat at will.

self-pollinated or *selfed.* Pollinated by transfer of pollen from stamen to pistil in the same flower.

septic. Causing infection or putrefaction.

shatter. To lose the leaves of grass before harvesting, or to drop grains prematurely.

sheaf. A bunch of cut stalks of grain bound together.

sheet erosion. A gradual washing away of topsoil in sheets by the action of wind or water.

shelter belting. The planting of one or more belts of trees facing prevailing winds.

shock. A bunch or pile of grain sheaves stacked in a field to cure and dry. A collection of cut cornstalks or wheat stalks. Also, a small hay pile.

shredder. A machine for tearing into small pieces the coarse stems of forage crops.

shrub. A bushy, woody plant of low stature with permanent stems instead of a single trunk.

sickle. A reaping implement with a long, curved blade mounted on a short handle.

silage. A fermentation product made from green fodder, preserved in a silo, and used as livestock feed.

silo. An airtight tower in which green fodder is stored.

silt. Any earthy material composed of fine particles, as soil or sand, suspended in or deposited by water.

silviculture. The art of cultivating a forest.

slip. A stem, root, twig, etc., cut or broken off a plant and used for planting or grafting; a cutting, a scion.

small grains. Small, hard seeds or seedlike fruits, especially those of any cereal plant, such as wheat, rice, corn, or rye.

sod. A surface layer of earth, particularly containing the matted roots of grass plants.

soil. Finely divided rock material at the surface of the earth, mixed with decayed vegetable or animal matter and supporting plant life.

soil-conserving crop. Any crop that maintains or improves the land.

soil-depleting crop. A crop that has a tendency to exhaust the land.

soiling. The act or practice of feeding cattle or horses with fresh grass or green food cut daily for them, instead of pasturing them.

soil structure. The arrangement of individual grains and aggregates that make up the soil.

sorghums. Coarse-leafed annual grasses with succulent stems, grown for forage, grain, or syrup.

sow. To scatter or plant seed in soil.

spawn. To produce or deposit, as a fish emitting a mass of eggs.

spike-tooth harrow. A farm implement used for smoothing and breaking topsoil.

splash erosion. Land leveling caused by raindrops.

sport. A plant or animal showing some marked variation from the normal type; a mutation.

spreader. A device for distributing manure.

springer. A young cow ready to calve.

spring wheat. Wheat planted in the spring and harvested in the summer of the same year.

sprout. A young growth on a plant, as a stem or branch; a shoot.

stack. A large pile of straw, hay, etc., especially one symmetrically arranged with a smooth outer surface for outdoor storage.

stalk cutter. A machine with which to overrun and cut dried stalks left in fields after crops are harvested.

stand. A standing growth of trees or plants.

standard. A freestanding tree or shrub, especially one not grafted onto another stock or supported or attached to a wall or trellis.

sterilize. To make unproductive or barren. To destroy germs or bacteria.

stolon. A shoot that proceeds from a stem above ground, takes root at the tip, and develops a new plant; a runner.

stoop crops. Crops that require many laborers to stoop to gather them.

stover. The residue of stalk, leaf, and shuck from grain crops after the ears have been removed, such as corn stover or milo stover. Used as winter provisions for livestock.

stratification. A process applied to certain seeds to speed up germination. Seeds are placed in layers of sand and exposed to the elements for one or more winters.

strip cropping. Alternating strips of heavy-rooted and loose-rooted plants to lessen erosion, as on a hillside.

strippings. The last milk drawn from a cow at a milking.

stubble field or *stubble pasture.* A field from which the grain has just been cut; a field covered with stubble.

subhumid. A climate with enough moisture to support a moderate growth of natural grasses but not enough to produce a dense hardwood forest.

subsoil. The bed or stratum of earth lying below the surface soil.

subsurface farming. Farming in which tillage tools break the subsoil but leave surface vegetation intact as mulch.

suckers. Shoots springing from the roots or the lower nodes of the main stem of a plant.

summer annuals. Plants that grow from seed each spring or summer and do not survive the winter such as soybeans.

swarm. The simultaneous emergence of a large number of bees from a hive, often to establish a new colony.

swath. The breadth or sweep covered with one cut of a scythe or mowing machine.

swill. Garbage, table scraps, etc., mixed with liquid and used as food for domestic animals, especially pigs.

tankage. Refuse from tanks, as the residue left in rendering out soap or fats; it is dried, ground, and used as a stock food or for fertilizing.

taproot. The main, central root of a plant which grows directly downward with small branch roots spreading out from it.

temporary pasture. Pasture used for grazing during a short period, not more than one crop season.

thinning. Reducing the number of plants to allow those remaining plenty of space in which to grow.

tillage. The use of implements to cultivate the land and to prepare seedbeds and rootbeds for crops.

top-dressing. The scattering of fertilizer or another material on land or crops without working it in.

topography. The science of drawing on maps and charts or otherwise representing the surface features of a region, including hills, valleys, rivers, lakes, canals, bridges, and roads.

topping. Trimming the top from a plant to increase the development of the remaining lower part.

topsoil. The uppermost layer of soil, usually darker and richer than the subsoil.

transpiration. The giving off of moisture through the pores of the skin or through the surface of the leaves and other parts of a plant.

trashy cultivation. Tillage that leaves the major portion of crop residues on the surface of the soil.

trench silo. A trench dug in the earth and used as a silo.

truck. Vegetables grown for market.

vertical erosion. The leaching of soluble soil nutrients down into the subterranean water table.

vetch. Any of a number of short, leafy, climbing or trailing, winter-growing legumes grown annually for fodder and as a soil restorer.

vitamin. A complex organic substance present in most foods and essential, in small amounts, for the normal functioning of the body.

wallow. A muddy or dusty place in which animals, such as the buffalo or swine, wallow.

warm-season grass. Grass that grows most luxuriantly during the warm season of the year.

water furrow. A deep furrow made for conducting surface water away and keeping the ground dry.

water-soluble. Able to dissolve in water; said especially of certain vitamins. Opposed to fat-soluble.

weed. Any undesired, uncultivated plant that grows in profusion and crowds out a desired crop.

whet. To sharpen by friction, as the edge of a tool.

whey. The thin, straw-colored, watery part of milk which separates from the thicker part (curds) after coagulation, as in making cheese.

whip or tongue grafting. A method of scion-grafting in which the scion and stock are approximately the same size.

wilt. A plant disease characterized chiefly by wilting leaves and caused by certain fungi or bacteria.

windbreak. A hedge, fence, or row of trees that serves as a protection from wind.

windrows. Sheaves of grain arranged in rows to let the wind blow between them; a long, low ridge of hay raked together to dry before being raked into cocks or heaps.

winter annual. A plant that germinates in summer or autumn, lives through winter, and matures and dies the following year.

winter wheat. Wheat sown in the fall and harvested the next summer.

Index

Page numbers in bold face refer to illustrations.

Jefferson, Thomas, 15

Kains, M. G., 29
Kansas Experiment Station, 99
King, F. H., 141, 153, 156, **158,**
    163, 281–82, 286–88, 293–
    94, 296
Kropotkin, Peter, 16

Lake States Forest Experiments,
    187
land
  availability of, 15–16
  building/zoning restrictions and,
    52
  contract of sale of, 54
  corporate control of, 16
    low-acre yield due to, 16
  cultivated, 15–16
    increase of, 16
  mortgage contract for, 54
  recording of sale of, 54
  sanitation requirements and, 52
  survey of, 52
  title search for, 52
legumes, 125, 141, 143, 158
ley crop farming, 125
lifting and carrying, **38**
light
  chlorophyll synthesis in plants
    and, 211
  color in, effects of, 210
    Pleasanton's discoveries about,
    211
  fluorescent, and photosynthesis,
    211–13
  incandescent, 212–13
  photosynthesis and, 202
  sun, source of, 201–2
    spectral qualities of, 210, **212**
  ultraviolet, 210
logging, 190

Malthusian theory, 15–16
manure, 163, 176, 291
  fish, 308
  spread by animals, 132
maple sugar, 194
map making, 59–63. See also
    homestead plan
Mercedes-Benz, Unimog, 236,
    238–39
methane gas production, 319–22,
    320
minerals
  in animal feed, 284
  in food, 334–35
  tablets, 348
Missouri Agricultural Research Sta-
    tion, 126, 161
modular gardening beds, 163
monoculture, 111, 157–58, 175
mowers, 149–51
mowing, 128–29
mulch, 92, 94, 109–10, 138–39,
    141, **155, 156,** 162–63, 172,
    175–76, 278, 291, 298
  defined, 93
  planting with, 127–28
  soil temperature and, **155**

National Health Federation, 349
Nearing, Scott and Helen, 32, 194,
    204
neohomesteading, 14, 15, 16, 25
Neubauer, Dr. Loren, 295
New Alchemists, 314
New Jersey Agricultural Experi-
    ment Station, 109
New South Wales Conservation and
    Irrigation Commission, 70
nitrates as fertilizer, 21
nitrification, 317–18
nitrogen, 124–26, 162, 188, 277
  cycle, **318**